PENGUIN BOOKS

DIATOMS TO DINOSAURS

Christopher McGowan is Curator of Palaeobiology at the Royal Ontario Museum in Toronto, and Professor of Zoology at the University of Toronto. He is the author of *Dinosaurs, Spitfires, and Sea Dragons*, *Make Your Own Dinosaur out of Chicken Bones* and *The Raptor and the Lamb*, which was recently published by Allen Lane and is forthcoming in Penguin.

DIATOMS

TO DINOSAURS

The Size and Scale of Living Things

Christopher McGowan

Illustrations by Julian Mulock

PENGUIN BOOKS

PENGUIN BOOKS

Published by the Penguin Group
Penguin Books Ltd, 27 Wrights Lane, London W8 5TZ, England
Penguin Putnam Inc., 375 Hudson Street, New York, New York 10014, USA
Penguin Books Australia Ltd, Ringwood, Victoria, Australia
Penguin Books Canada Ltd, 10 Alcorn Avenue, Toronto, Ontario, Canada M4V 3B2
Penguin Books (NZ) Ltd, Private Bag 102902, NSMC, Auckland, New Zealand

Penguin Books Ltd, Registered Offices: Harmondsworth, Middlesex, England

First published in the USA by Island Press 1994
First published in Great Britain in Penguin Books 1999
1 3 5 7 9 10 8 6 4 2

Printed in England by Clays Ltd, St Ives plc

To Claire and Angie—daughters who make a father proud.

CONTENTS

CONTENTS

ACKNOWLEDGMENTS

IF THIS BOOK has merit it is because of the excellent research being conducted in laboratories around the world and reported in the literature. Many of the authorities of this rich resource have taken the time and trouble to read individual chapters of the first draft, and it is my privilege to acknowledge them first. I am deeply moved by their generosity in critically reading the manuscript and making so many valuable comments—in many instances even sharing their own unpublished manuscripts with me. It is an honor to be in such fine company. My sincere thanks to R. McN. Alexander, R. W. Blake, C. P. Ellington, C. E. Finch, J. W. Gibbons, J. L. Gittleman, J. B. Graham, J. Hanken, P. H. Harvey, A. M. Kuethe, J. Machin, J. H. Marden, R. MacDougall, B. W. McBride, W. Nachtigall, D. Norman, M. D. Pagel, R. H. Peters, D. E. L. Promislow, E. M. Purcell, R. S. Seymour, M. A. Sleigh,

D. Stevens, C. R. Taylor, M. Thomas, J. J. Thomason, S. Vogel, C. S. Wardle, and P. W. Webb.

I am remarkably fortunate to work at a museum that still values the importance of research and scholarship. I am also fortunate to have a cross-appointment to the Department of Zoology of the University of Toronto. I have the best of both worlds—collections to stimulate the curiosity and students to challenge my interpretations of them. I wish to thank both institutions for their support over the years, which has enabled me to pursue my interests in the living world. My thanks also go to a number of colleagues who have been especially helpful during the present project: Allan Baker for discussions on various aspects of evolution and molecular biology; Dale Calder for supplying invertebrate literature; Chris Darling for sharing his entomological knowledge and for directing me to the relevant literature; John Machin for enlightening discussions on things functional; T. Sawa for assistance with the literature on diatom locomotion; Berry Smith for information on *Trichogramma*; and Sandy Smith for supplying live specimens of these intriguing little wasps. For access to specimens in their respective departments I thank Jim Dick, Erling Holm, Burton Lim, and Marty Rouse. Thanks to Catherine Skrabec for her numerous trips to the library, literature searches, xeroxing, clerical help, and for unraveling my computer problems. Julia Matthews and her staff at the ROM library patiently searched files and arranged interlibrary loans, for which I am grateful. Thanks also to Mindy Myers for so many book-laden sorties to and from the University of Toronto library. I thank Brian Boyle and Ron Pozniak for photography. I am sincerely grateful to Rosemary MacDougall for reading most of the first draft, giving encouragement and many valuable criticisms, and for entering corrections at a time when help was most needed.

Working with Julian Mulock was the same great pleasure it has always been. I thank him for his outstanding artwork, his sympathy for the subject, his enthusiasm for the project, and for his good humor.

The idea for writing a book on size was Howard Boyer's, my editor—we both wanted an excuse for working on another project together. As on the previous occasion, this has been a rewarding and enjoyable experience. I thank Howard for his enthusiasm, encouragement, and for his honesty; those comments seared onto the pages of the first draft made for a much improved manuscript. No author could wish for a more competent and speedy passage of a manuscript than I have received at Island Press and I am sincerely grateful to them for making it all happen so efficiently and smoothly. My special thanks to Sam Dorrance for his continued interest in the book, for his zeal in seeing it through to completion, and, most important, for getting it into the bookshops. Thanks also to Siobhan White for her thoughtful attention to so many details, including the reviews. I also thank J. P. Long, also of Island Press. Susanna Tadlock and her associates took care of production, Nancy Evans of Wilsted & Taylor edited the final draft, and Dave Bullen produced the design. I thank each one of them for a superb job, completed with care and attention during a remarkably tight schedule. Unheard-of books are seldom read and I am fortunate to have had Mary Bisbee-Beek as publicist. I thank her for her enthusiastic interest and dynamic promotion of the book.

Liz McGowan proofread the entire manuscript—twice—put up with my spending every evening and weekend in front of the computer, offered valuable opinions and constructive criticism when asked, and gave her support, encouragement, and love all the time. I count my blessings.

The Natural Sciences and Engineering Research Council of Canada has funded my research for almost two decades, for which I express my thanks and gratitude.

DIATOMS TO DINOSAURS

THE SCALE OF LIFE

THE LAND TO the west still slumbers in the indigo world of night, but the eastern sky has already taken on the sienna tones of dawn. Acacia trees, their umbrella crowns silhouetted against the reddening sky, dot the horizon like lone sentinels, awaiting the start of another day. A lioness roars from a long way off, adding her distinctive voice to the night song of insects. A herd of elephants shuffles through the shadows, a legion of gargantuan insomniacs. Their trunks probe here and pull off a branch there as they browse their unhurried way across the land. Although their metabolic furnace does not burn as brightly as that of a smaller animal, their immense size requires that it be kept well stoked. They spend at least eighteen hours of every day feeding, stealing a few fitful hours of sleep while standing up. Sometimes they lie down to sleep for an hour or so, and then they must arise, otherwise their crushing weight would cause permanent or even fatal tissue damage.

Daybreak. The African sun rises majestically into the heavens, spilling light and warmth across the land. As the ground begins to warm, weak convection currents stir in the air. Some parts of the land heat more rapidly than others, spawning thermals. At first the thermals lack the strength to carry anything heavier than the smallest objects heavenward—a stray feather, a small fly, a strand of spider's silk. But as the sun gets hotter and the thermals grow stronger they are able to support the soaring flight of birds. The first to take to the air are the smaller ones, like hawks, whose wings have the lightest loads to bear. The larger and heavier vultures have to wait for stronger currents to carry them aloft.

High noon. A heat haze shimmers across the parched savannah. A bull elephant stands beneath a tree, the dappled shade barely covering his body. His huge bulk makes him particularly vulnerable to heat stress, and partial shade is better than no shade at all. The elephant raises his trunk like a periscope and sniffs the air. Among the familiar scents of wildebeest, zebra, and gazelle, he picks out the scent of lion. Not that this is of any great concern—the elephant is forty times heavier than the lion, a size advantage that makes him invulnerable to attack. The trunk swivels to the south and the elephant takes another long, contemplative sniff. This time he finds what he is searching for—water, but a long way off. The elephant now catches sight of a solitary lioness, creeping stealthily through the long grass toward a herd of unsuspecting zebra. A seasoned hunter, she makes her approach from downwind, freezing whenever one of the zebras looks up from grazing. Zebras can run faster, but lions, with their powerfully muscled limbs, have a greater acceleration. If the lioness can get close enough to the zebras without being seen, she stands a chance of making a successful charge. Although most attacks by solitary lions end in failure, the lioness will be lucky today.

Late afternoon. The lioness, having eaten her fill, dozes contentedly beside her kill. The rest of her pride, muzzles bloodied and bellies full, are stretched out in the sun beside her. A vulture, black-winged and bare-necked, swoops down on the carcass for a free meal. Its bald head disappears inside the zebra's abdominal cavity. But it has barely got a beakful of offal before it is rudely swatted away by the lioness. She is unwilling to share the smallest scrap of her prize with the vulture, even though armies of insects have already set to work devouring the carcass. They are big enough to be seen by the lioness, but far too small to concern her. But their small size belies their enormous strength; on a pound-for-pound basis they are far stronger than the mighty lion. No lion could lift an object weighing several times its own body mass, but this is routine business for insects.

Other insects, like the flesh flies that swarm everywhere, are also interested in the meat—not for themselves but to sustain the larvae that will hatch from the eggs they are busy laying in the zebra's flesh. Still others, like the dermestid beetles, have different tastes; when their larvae hatch they will nibble away at the hide, ligaments, and tendons. These voracious scavengers, though less than 1 inch (2.5 centimeters) long, will strip the carcass to bare bones within a week or two.

Early evening. Heavy rain clouds gather in the dark sky like mourners at a funeral. A few large spots of rain fall to the earth but they are instantly devoured by the thirsty soil. It is too early in the season for rain, and this token display will not amount to anything. One or two droplets fall on the massive flanks of an elephant, spreading ochre stains on its dusty gray hide. The great beast does not feel the rain and continues pushing against a large tree with its massive head. There is nothing hurried in its action, just the patient application of an irresistible force to a seemingly immovable object. Moments later the tree keels over. An exploratory trunk probes the freshly exposed tree roots—a fastidious diner perusing the menu—then the feeding begins.

A small mound of twigs and dried leaves no higher than a golf ball twitches inexplicably on the dusty ground. Then a small head appears—pointed snout, big ears, long whiskers—and the mystery is resolved. The tiny rodent scampers nervously from its temporary concealment and sets off in search of another place to find food. So far its search has been successful and it has found a good crop of seeds. But its high metabolic rate requires a large intake of food so it must keep on with its frenetic search for more. It scurries across the bare ground—whiskers twitching, eyes peering—alert to the ever-present threat of attack. The only defense for such a small and vulnerable animal is concealment, and in its haste to find cover it almost collides with a tree. The width of the tree is four times the rodent's length—the equivalent of a human encountering a house. The small creature stops in its tracks as if deciding on which side to pass, but before it has made up its mind the tree disappears. The rodent scurries on, oblivious of the fact that its short life almost ended beneath the foot of an elephant.

Another dawn. The rain clouds have moved off to the north and the coral sky awaits the sun. A small herd of alert impala graze within sight of the lions. But the lions are too engrossed in finishing off the last of the zebra to pay them any heed. A loop of intestine, grossly distended with gas, protrudes from the carnage like an obscene balloon. The gas is the product of bacteria, organisms that live by the billions wherever there is moisture. They are far too small to be seen without a microscope but their accomplishments are monumental. The smell of putrefaction is their doing, and this will ripen into an unbearable stench by day's end. One of the cubs takes a playful swipe at the bloated intestine, which bursts with a hiss of putrid gas. He gets a swat from the lioness for his trouble.

The lions, the insects, and the bacteria live in separate worlds on separate scales, but their lives are intimately entwined. If the lioness had not killed the zebra there would be no temporary universe for the scavenging insects and the putrefying bacteria. And if there were no scavengers and putrefiers to break down dead animals, no animal nutrients would return to the soil; hence there would be fewer plants, fewer zebras, fewer lions.

Bacteria are the smallest of organisms and are measured in thousandths of a millimeter, or micrometers (abbreviated μm). The largest organism on earth, the blue whale, can grow to more than 100 feet (31 meters), 10 million times the length of a bacterium. While organisms are set worlds apart from each other by these enormous differences in scale, they are all governed by the same physical principles. One of these principles, the relation between length, surface area, and volume, plays an all-pervading role in living things, just as it does in the inanimate world. We see examples of it every day of our lives, but we either ignore them completely, or take them for granted.

Saturday morning. No alarm clock ringing, no rush to get up. You take a leisurely stroll to the kitchen, make some coffee, and pop a muffin into the microwave. Two minutes later you have your coffee, the morning paper, and a sizzling muffin that's too hot to eat. You cut the muffin into slices to let it cool, turn to the sports section, and carry on reading. Even if you had stopped to think about it, I doubt you would have consciously reasoned that in cutting the muffin into smaller pieces you were increasing its surface area relative to its volume. Since most of the heat escapes from the hot muffin through the surface, increasing its area-to-volume ratio speeds up the cooling.

Still feeling hungry after eating the muffin, you contemplate the donuts left over from last night. Ignoring the chocolate and the plain ones, you choose the one with the powdery white dusting. The instant your tongue touches it, the dusting dissolves in a gratifying burst of sweetness. They must use some special sort of sugar at the bakery because the stuff you put in your coffee tastes nowhere near as sweet. Wrong. It is the same sort of sugar but is just more finely ground. Being so small, the sugar particles have a huge area-to-volume ratio, so they dissolve more rapidly, giving the impression of greater sweetness.

While you lick your fingers and abstemiously close the lid on the donuts, the marmalade cat on the sofa stretches, yawns, and goes back to sleep. You glance in her direction and notice how quickly she is breathing—and she is not even doing anything. If you checked her pulse you would find that her heart was beating almost two hundred times a minute compared with your rate of sixty-five. The fact that small animals have faster breathing rates and heart rates than large ones is another consequence of differences in area-to-volume ratios.

The heart has to pump blood around the body through a vast network of blood vessels. Most resistance to the flow comes from the narrowest blood vessels, the capillaries. Capillaries supply blood to all parts of the body, mostly near surfaces such as the lining of the lungs, gut, and skin. Smaller animals have larger surface areas for their size, so the resistance to the passage of blood rises with declining body size. There are two options for coping with this problem—larger hearts or faster heart rates. Animals usually adopt the latter strategy and the size of the heart remains fairly constant relative to the size of the body. For mammals the heart weighs an average of 0.6 percent of the body weight. The heart rate of the smallest mammals, the shrews, which only weigh 3 grams, can reach a staggering 1300 beats per minute during activity. Hummingbirds similarly have small bodies and high heart rates.

You are on your second cup of coffee before the rest of the family wakes up. The four-year-old is the first one downstairs. He hauls himself up on your lap and reaches for the donuts. He is not heavy—only 40 pounds—but what a fidget! If you stopped to think about it, you might expect him to weigh far more than he does; he is, after all, half your height, and you weigh 190 pounds—almost five times heavier. If your son were an exact, half-size replica of yourself, he would weigh eight times less than you do because weights scale with the cube of the difference in length. This is best shown for simple shapes like cubes, as depicted in the illustration (p. 10). You can also see why area-to-volume ratios decrease with increasing size, and how these variables can be shown graphically. The relation between length, area, and volume is a recurrent theme so it is worth taking the time to review this material.

Since living things have bodies made of similar materials, mostly water, they have similar densities, so we can use volume and mass interchangeably. Therefore if an animal's body had three times as much volume as a smaller one, it would also be three times heavier.

If offspring were exact replicas of their parents, then photographs taken of each and printed to the same size would match when superimposed. They do not match, though, because body proportions change with growth. One of the most obvious differences is in head size—babies proportionately have much larger heads than adults. This change in proportion as animals grow larger has been recognized for over a hundred years. It was not until the first half of the twentieth century, however, that the subject of *allometry* (Greek, *allios*, different; *metron*, a measure) was treated mathematically. The central figure in these studies was the English biologist Julian Huxley, grandson of Thomas Henry Huxley, staunch defender of Darwinism. In a book published in 1932 Huxley showed that a diverse assortment of growth phenomena, ranging from fiddler crabs' claws to deer's antlers, could be summarized by a simple equation. The equation is $y=bx^k$, where y is organ size (say diameter of the head), x is body size (such as body length), and b and k are constants. Equations of this kind, where one variable changes according to another variable raised to a power, are called power functions, and the power term, k, is called the exponent. If you measured head diameters and body lengths for a series of humans, from newborn babies to adults, and plotted a graph of head size (along the y-axis) against body size (along the x-axis) you would get a convexly curved graph. It would be a fairly simple matter to enter the x and y data into a computer, generate a graph, and calculate the values b and k. But there were no computers in Huxley's time, not even pocket calculators, so he needed some simple way of solving the equation. His solution was elegantly simple. All he did was plot the logarithms of the variables instead of the variables themselves. Without going into the mathematics, this gives a straight-line graph instead of a curved one, and the slope or gradient of the graph, easily measured using a ruler, is the exponent, k. The other constant, b, can be read from the graph where the line cuts the y-axis. If you plotted logarithmic data for head and body size for humans (by looking the number up in a set of log tables) you would get a straight-line graph whose slope was about 0.7. Exponents like this of less than 1 mean that as the animal gets bigger the organ gets smaller relative to the body. Such growth is called *negative allometry*. Sometimes the exponent has a value of 1, meaning that the growth of the organ keeps pace with that of the body. The proportions therefore do not change and this is referred to as *isometric growth* (Greek, *isos*, equal). Huxley's fiddler crabs provide examples of both isometric growth and *positive allometry*, where the exponent is more than 1. Small fiddler crabs have fairly small claws, and in females their growth is isometric so they remain proportionately in scale with the rest of the body as it gets bigger. The same is true for one of the male's claws, but the other one grows disproportionately large, until it is almost as big as the rest of the body. The strong

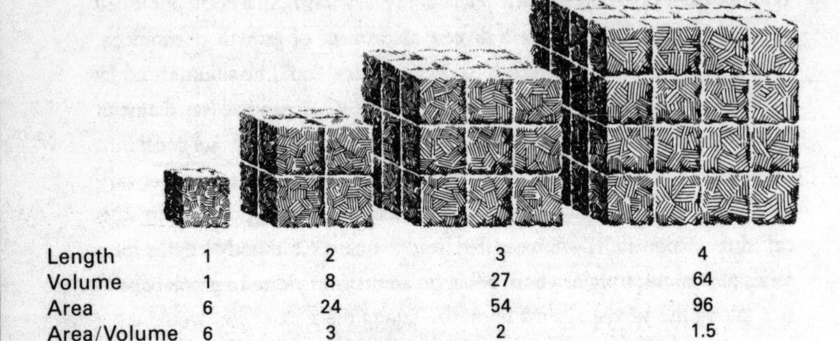

Length	1	2	3	4
Volume	1	8	27	64
Area	6	24	54	96
Area/Volume	6	3	2	1.5

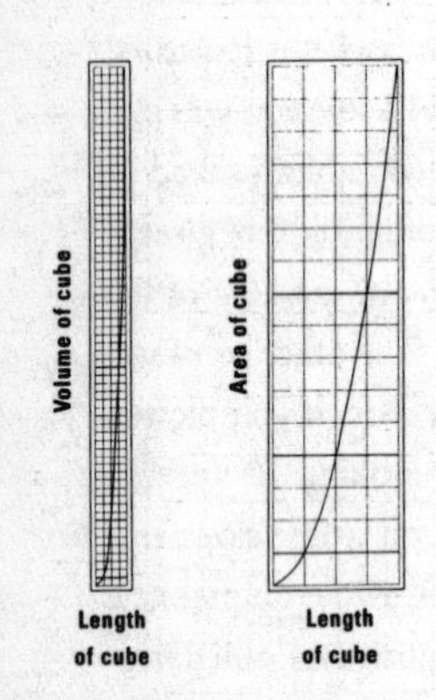

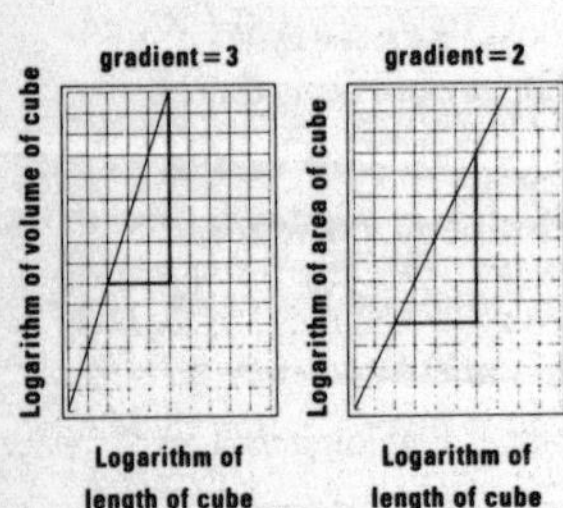

As things get bigger their volumes, hence their masses, increase very rapidly, but their areas lag behind. This is best illustrated with simple shapes like cubes.

Compared with the first cube, the others are 2, 3, and 4 times longer. Their volumes—the numbers of cubes they contain—are 1, 8, 27, and 64 respectively. Compare the first two cubes. Doubling the length increases the volume by 8, that is by (length increase)^3. Compare the first and third cubes. Tripling the length increases the volume by 27, that is by (length increase)^3. Quadrupling the length from the first to the fourth cubes increases the volume by 64—this again is by (length increase)^3. Thus the volumes, hence the masses, increase with the cube of the length increase.

Each cube has six faces so their surface areas are six times the area of one face. The areas of the four cubes are therefore 6, 24, 54, and 96. The area-to-volume ratios are therefore 6/1 = 6, 24/8 = 3, 54/27 = 2 and 96/64 = 1.5.

*In the graphs at left, when volume is plotted against length a steeply curving graph results (*top left*). If logarithmic rather than raw data are plotted, a straight line graph results. The gradient of the graph is the exponent, 3 (*bottom left*).*

*Areas increase less rapidly than volumes. So when the areas of the cubes are plotted against their lengths the graph is less steep than it was for their volumes (*top right*). When logarithmic data are plotted a straight line results whose gradient is 2 (*bottom right*).*

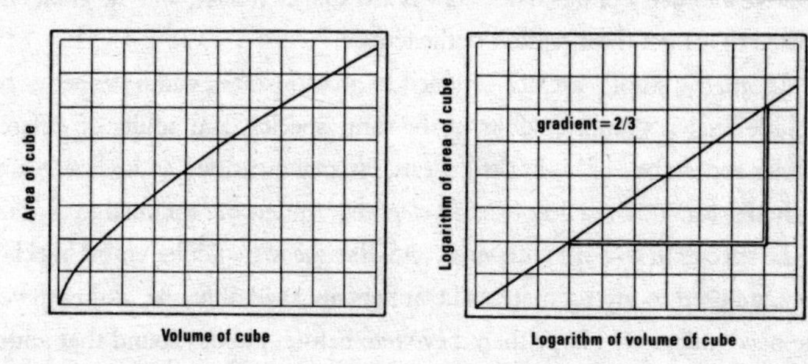

Areas decrease relative to volumes. When the areas of the cubes are plotted against their volumes a gentle convex curve results (left). *When logarithmic data are plotted a straight line results whose gradient is 2/3, the exponent. Whenever logarithmic graphs of biological data give gradients of 2/3 area-to-volume relations should be suspected.*

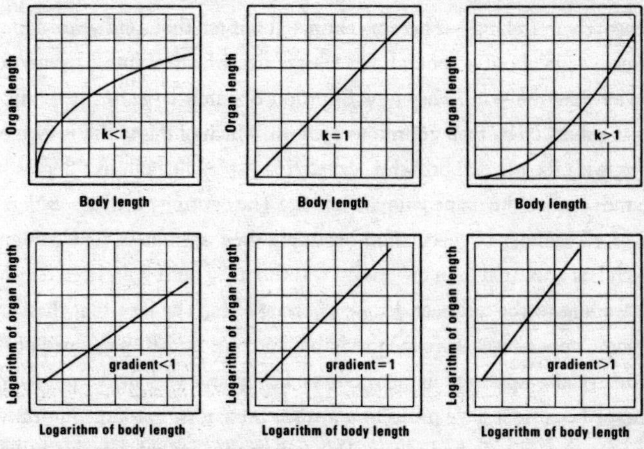

Allometric growth occurs when different parts of a body grow at different rates to the rest. Parts that grow allometrically give curved graphs when plotted against body size (top row). *When logarithmic data are plotted instead of raw data, straight-line graphs result* (bottom row). *When the growth constant,* k, *is less than 1 (negative allometry), the curve is convex* (top left); *when it is greater than 1 (positive allometry), the curve is concave* (top right). *If the body part grows at the same rate as the rest of the body, the growth is called isometric. Parts that grow isometrically give straight-line graphs when plotted against body size* (top middle). *Logarithmic graphs have gradients of less than 1 for negative allometry* (bottom left), *greater than 1 for positive allometry* (bottom right), *and 1 for isometry* (bottom middle).

positive allometry of the male's claw is a sexual character, and the giant claw is used to attract the attention of the female.

Allometric studies are not confined to growth stages within a species but include analyses among adults of the same species, and adults of different species too. In one of his investigations, Huxley measured antler size against body size for adult red deer (*Cervus elephus*). Antlers are confined to the male of the species in true deer (cervids), and, like the male fiddler crabs' big claw, they are used to attract mates. The antlers are shed annually, and each year the new antlers are bigger than they were before. Huxley found that antlers had a positive allometry, becoming relatively larger with increasing body size.

During the Ice Age there lived a giant deer with a shoulder height of about 6 feet (2 meters) and antlers up to 12 feet (4 meters) across! Inappropriately named the Irish elk (*Megaloceros giganteus*)—it is neither an elk nor found exclusively in Ireland—it became extinct about ten thousand years ago. The antlers weigh about 90 pounds (41 kilograms), which is almost twenty times heavier than the skull. They look disproportionately large for the body, and this is enhanced by their enormous breadth. Much of this width is formed of bone, with the pointed branches, called tines, spreading out like fingers from a hand—hence the name palmate antlers. The enormity of these antlers has led to all manner of speculation regarding their usefulness. The consensus, which has appeared in many books, was that they must have been detrimental, leading to the ultimate demise of the species. The idea that these cumbersome appendages were the inevitable outcome of their presumed positive allometry also appeared in many books: being large was an evolutionary advantage because it gave protection against predators, but large antlers were the price the Irish elk had to pay. This notion that animals become locked in to growth constraints fails to consider how species evolve.

Harvard University's Stephen Jay Gould, whose insightful mind and brilliant writings have done so much to popularize biology, got interested in the Irish elk in the early 1970s. Until that time none of the speculations about the beast had been supported by any hard evidence. Gould therefore set to work and measured as many specimens as he could, in museum collections on both sides of the Atlantic. His studies showed that the Irish elk's antlers, like those of the red deer, did indeed have a positive allometry. When he plotted a logarithmic graph of antler size against body size for various deer species, the Irish elk fitted on the line, showing that its antlers were the "right size" for its large body.

What was the purpose of such large antlers? Offense and defense can be ruled out as most unlikely. As Gould noted, the full extent of the Irish elk's antlers are seen when the head is directed forward, as in a displaying stance, rather than when the head is lowered in an attacking posture. The moose, the closest living analogue of the Irish elk, uses its feet rather than its palmate antlers for warding off attackers, and its antlers are mostly used for attracting mates and intimidating other males. It therefore seems likely that the Irish elk's antlers had similarly evolved for display purposes, and were probably every bit as "useful" as the elk's large body size. The antlers should therefore be viewed as structures that served a useful function and which were the appropriate size for a deer of such large proportions.

Phylogeny (Greek, *phyle*, race; *genos*, birth, descent), the study of relationships among organisms, has made considerable progress during the last two decades. Part of this progress is attributable to the availability of new techniques for assessing such relationships. One of the most powerful of these tools is DNA analysis, where the triplet codes locked up in the genes are revealed in the laboratory. DNA sequencing, for example, has revealed that our closest relative among the primates is the chimpanzee rather than any other of the great apes. The objective of phylogenetics is to establish natural groupings of organisms, that is, a group of organisms that all evolved from a common ancestor. Insects, for instance, form a natural or monophyletic (Greek, *monos*, single) group, and it is believed that they all evolved from a common ancestor. Descent from a common ancestor is established by seeking specialized, or derived, features that all members of the group share. Insects share a number of derived features, most of them pertaining to details of their internal anatomy. Included in these features are the possession of an ovipositor, a structure used for depositing eggs, and valves on the spiracles, the holes through which insects breathe. The respiratory system of insects comprises a network of fine ducts (tracheae) that deliver air directly to all parts of the body. Although the system works well on a small scale, it cannot be expanded to serve the respiratory needs of large animals. The size constraint imposed by the tracheal system explains why insects have been unable to exploit the advantages of gigantism during their long evolutionary history. Most flying insects, for example, are specialists at the small end of the size spectrum—no insect aeronauts compare in size with the giant pterosaurs. Like any other group of organisms, insects are constrained in the evolutionary pathways they can follow by the phylogenetic baggage they carry. I will not belabor this in the pages that follow, but it would be good to bear this in mind.

Huxley's classical treatment of size seeded a veritable growth industry in allometric studies, relating body size to everything from brain size and metabolic rate to life span and population density. Many of the studies were entirely descriptive, and attempts to explain the underlying biological phenomena have often been unsatisfactory. This is especially true of explanations that were not made within an evolutionary framework of natural selection. The idea that the Irish elk's large antlers were an inescapable consequence of their positive allometry, for example, sidesteps the question of how large antlers may have evolved within the species. That is not to suggest that we have all the answers today—far from it. But we are more likely to approach an understanding of what is going on if we view things in an evolutionary light.

Consider the following facts: A mouse weighs about 1 ounce (about 20 grams), has 700 heartbeats a minute, a gestation period of 21 days, has multiple births, a brisk metabolic rate, and lives for about three years. A five-ton* elephant has 30 heartbeats a minute, a 22-month gestation period, bears single offspring, has a slow metabolic rate, and lives for sixty years or more. All manner of proposals have been made linking some or all of these facts together, but we can only begin to make sense of them if we tackle the problem from an evolutionary perspective.

Our sojourn in the African savannah is just a taste of things to come. In the pages that follow I hope to share the excitement of discovery and the joy of understanding, and by journey's end we will have a clearer view of the

*English tons and metric tons are almost the same and I will not distinguish between them in this book.

biological world in which we live. Size influences all aspects of living organisms and provides an intriguing perspective from which to view life. Owing to the constraints of the size of a book, I have limited my subject matter primarily to whole organisms, rather than considering the internal workings of their parts. Because of my zoological background, I focus almost entirely on animals, mostly those with backbones. Some are chosen from the Mesozoic Era, the Age of Reptiles. Dinosaurs, the largest animals ever to walk the earth, will provide the most spectacular examples of gigantism. For the minutiae of life, I will explore the sea and the captivating world of plankton.

Any organization of a subject that transcends so many areas of biology is bound to be unsatisfactory, but a random collection of topics would be more undesirable. Since animals living on land are subject to a suite of forces different from those living in water or flying in the air, I have organized the material of the book into three categories: land, water, and air. Inevitably some topics spill over these artificial boundaries. For instance, the thermal implications of large size affect whales and elephants in the same way, but most of my examples of thermal biology pertain to life on land, and so I deal with the subject there.

Living organisms are far more complex and more difficult to study than nonliving systems, which is why inanimate systems are much easier to understand. An aircraft wing may seem discouragingly complex from the passenger seat—especially during takeoff and landing when an array of flaps and slots are deployed—but it is a simple device compared with a bird's wing. To elucidate certain principles about organisms, then, I will draw freely from the physical world. The first such principle I want to address is metabolism, and this is the subject of the next chapter.

KEEP THE HOME FIRES BURNING

ONE SUMMER A few years ago, I was digging for marine reptiles in southern Manitoba. The prairies are not noted for their hills, but our group was in prime farmland, set in rolling countryside. We were camped on a farm, and our congenial host would drop by every day, usually with a pail of fresh milk still warm from the cow. We were as interested in things agricultural as he was in things paleontological, and when he asked if we would help a neighbor round up some chickens for market, we jumped at the request. The plan was to wait until evening, when the birds were settling down for the night, and then creep into the henhouse, grab the unsuspecting chickens, and whisk them off to the waiting truck.

We had a blissfully cool night for the roundup, but when we stepped inside the henhouse to do the dastardly deed, it was as hot as an oven. Chickens produce lots of heat, especially when several hundred of them are crowded into a low-roofed building. Birds, like mammals, are warm-blooded, meaning that they maintain high and fairly constant body temperatures, day and night. Mammals, ourselves included, maintain temperatures of around 98° F or 37° C; birds run a few degrees higher. This temperature is independent of climate—whether it be tropical or polar. Significantly, small animals generate more heat relative to their body mass than large ones, so the henhouse would have been even hotter if it had contained the same total mass of sparrows, but cooler if stocked with the same total mass of ostriches. To appreciate the difference in scale, consider that a 900-pound (409-kilogram) resting camel would generate heat equivalent to four 60-watt light bulbs, whereas 900 pounds of mice would generate heat equivalent to about sixty-four light bulbs. Where does all this heat come from?

All the cells of a body generate small amounts of heat as a by-product of the chemical processes going on inside them. This activity is called *metabolism*. Different cells metabolize at different rates, and therefore contribute different amounts of heat to an animal's heat budget. In a vertebrate body, among the most metabolically active cells are those of the kidney, liver, gut, and brain. These calls generate the most heat, and, although their respective organs constitute a relatively small percentage of the total body mass, they contribute a significant portion to the total heat budget. For example, in chickens the liver, gut, and reproductive organs together make up only 5–6 percent of the body mass, yet they account for 26 percent of the heat budget.

Cells with the highest metabolic rates not only generate the most heat but also consume the most oxygen. Oxidation takes place inside mitochondria—minute oval-shaped structures inside the cell. The site of this chemical activity is the mitochondrion's internal surface. While this surface is already large because of the mitochondrion's small size, the surface area is further increased by a series of internal folds that project inward from its lining. The total surface area provided by mitochondria for oxidation within cells is huge. Since the mitochondrion is the site of oxidation it follows that the most metabolically active cells have the highest densities of mitochondria. A cell's mitochondrial density therefore gives a good idea of its metabolic rate.

Metabolic rates vary not only among species, and from one individual to another, but also within an animal, depending on the animal's activity. While you are sitting down reading this book your metabolic rate is low, but if you got up and took your dog for a walk your rate would rise, and would increase even more if you started running. More heat is produced and more oxygen is needed as your metabolic rate climbs. When you are asleep your metabolic rate falls to its lowest level. But it will not fall so low if you have a large meal before going to bed, because that increases the metabolic levels of all the organs associated with digestion: stomach, intestine, liver, and pancreas, together with the heart and lungs, which are responding to the demand for more oxygen. So when your mother used to tell you not to eat a big meal before going to bed, it was good advice. Who wants to work during their sleep?

Since metabolic rates are so variable, a method has to be devised to compare different individuals and different species. It is usually easier to measure oxygen consumption than heat output, and this is done by measuring how much oxygen is consumed in a given time. The individual being measured has to be at rest, has to be fasting, and has to be within a comfortable temperature range. The last condition is required because changing ambient temperature can alter metabolic rates. This applies to both warm-blooded and cold-blooded animals, though the two groups react differently. If it gets cold our metabolic rate rises to produce more heat. We know by experience that if ambient temperatures are really low we start shivering, which is a mechanism for generating extra body heat. If the ambient temperature is too high, we start sweating to lose excess heat, an activity that also raises our metabolic rate.

The temperature range in which we feel comfortable—neither too hot nor too cold—is called the *thermal neutral zone*, the range in which metabolic rates stabilize. Metabolic rates for warm-blooded animals are therefore measured when the resting, fasting individual is within its thermal neutral zone. When measured under these conditions the metabolic rate is referred to as the *basal metabolic rate*, much the way that your doctor attempts to measure your blood pressure while you are sitting calmly in her office. Cold-blooded animals do not have a thermal neutral zone, but most have a preferred temperature range—the temperatures they would select if given a choice—and their basal metabolic rate is therefore measured within this range. Cold-blooded animals do not shiver or start running about to keep warm when the ambient temperature falls. Instead, they may try and find a warmer location, but if that fails their bodily functions just slow down; that is, their metabolic rate decreases. The reverse happens when ambient temperatures rise (since their metabolic response differs from that of warm-blooded animals, it is usual to state the temperature at which their basal metabolic rates are measured). The single reptilian exception to all of this is the brooding python. When female pythons that are incubating eggs are exposed to lowered ambient temperatures, they elevate their metabolic rates by muscular contractions, thereby enabling them to maintain a high (about 93° F, 34° C) and stable body temperature to keep the eggs warm. This response is limited, though: pythons cannot cope with ambient temperatures lower than 79° F (26° C), and it happens only with females that are with eggs.

When I was a student I had a part-time job working for a company that supplied schools with laboratory animals such as rabbits, frogs, dogfish, and crayfish. The dogfish were supplied to the schools pickled in formalin, but the others were usually requested alive, so we had to take care of them while they were waiting for shipment. The rabbits had to be fed and watered every day. Well, that was easy enough because they ate food pellets, with an occasional lettuce leaf. But how to feed a frog or a crayfish? When I started working for the company I had visions of patiently offering mealworms to each frog. And I had no idea what to feed the crayfish! But there was no need for concern—the frogs and crayfish were kept in the refrigerator. Being warm-blooded I would not have fancied it much myself, but they were cold-blooded and seemed quite happy with their accommodations. The reason for keeping their bodies cool was to lower their metabolic rates to the point where they did not need to be fed during their short stay with us.

The terms warm- and cold-blooded are in everyday usage, but they do not tell us very much about an animal's internal workings. Mammals and birds, by virtue of their high metabolic rates and body insulation (fur and feathers), can maintain high and constant body temperatures, regardless of ambient temperatures and their activity levels. Our cat's body temperature is about the same whether she is asleep in the shade or running about in the sun, summer or winter. To describe her warm-bloodedness more precisely, we would say that her thermal strategy is one of *endothermy* ("heat from within"). We might also use the term *homeothermy* ("equal heat"), which emphasizes the constancy of her body temperature rather than the source of heat, but endothermy will be used here. I noted that endotherms maintain constant body temperatures, day and night, but this is not strictly true because their temperatures usually drop by a fraction of a degree when they are asleep. In some mammals and birds, especially the smallest ones, the nightly fall in temperature is a few or several degrees. Roosting hummingbirds and bats may even allow their body temperatures to fall close to ambient temperatures at night, and hibernating animals, like ground squirrels, do the same thing during their winter sleep. This strategy, which reduces energy requirements during periods of inactivity, is sometimes described as *heterothermy* ("different heat"), though many prefer to use "endothermy." Being endothermic is an expensive business in relation to food requirements, so why should mammals and birds go to the trouble and expense of maintaining such high body temperatures? One of the primary advantages is probably an independence from the environment. Unlike cold-blooded animals, endotherms are always ready for action, regardless of ambient temperatures. It has also been suggested that enhanced muscle performance may be a primary factor. Certainly warm muscles generate more power than cold ones, a fact well known to athletes.

Birds and mammals are the most metabolically active animals and their basal metabolic rates are about ten times higher than those of similarly sized reptiles, amphibians, and fishes, as well as other multicellular animals, like insects and mollusks. These, in turn, have metabolic rates that are ten times or more higher than those of unicellular organisms, like amoebas. A twenty-gram lizard, for example, generates about 0.5 watts per kilogram of body mass at rest, compared with about 10 watts for a mammal of similar size. Because reptiles have much lower metabolic rates than mammals and birds, and lack fur or feathers, they are unable to maintain high and constant body temperatures endothermically. Yet most reptiles do manage to maintain high body temperatures of about 95° F (35° C) during the daytime, which they achieve by basking in the sun. At daybreak their bodies are cold and their movements sluggish, but after sunning themselves their metabolic rates and activity levels increase, as anyone knows who has tried catching a lizard in the sun. When they get too hot they seek the shade and return to the sun when their body temperature drops below the optimum level. This strategy of deriving energy from the sun is described as *ectothermy* ("outside heat") and gives reptiles the advantage of having a warm body without the high food costs associated with endothermy.

Just how economical ectothermy is compared with endothermy was illustrated for me when we kept a small caiman as a pet. Cedric, who was about 1 foot long (30 centimeters), used to be fed an ounce of liver once a week, but our cat demands three meals a day! The downside of being ectothermic, though, is a reliance on the sun, which obliges most reptiles to be inactive at night and during cold periods.

While mammals, birds, and reptiles exercise varying degrees of control over their body temperatures, the majority of animals have no control other than seeking more comfortable locations, such as under stones. These animals are at the mercy of the environment and their body temperatures fluctuate with ambient temperatures. Such a thermal strategy, or rather lack of one, is referred to as *poikilothermy* ("various heat"). Most animals are poikilotherms, including amphibians, most fishes, and invertebrates (with some fascinating exceptions, as we will find later). Because humans and their domestic animals are endothermic, most research on thermal strategies has focused on mammals and birds. These investigations date back to the nineteenth century, and it became clear early on that larger animals had relatively lower metabolic rates than smaller ones. The pioneer work in the field, by the eminent physiologist Max Kleiber, was a survey of the metabolic rates of mammals ranging in size from rats to cows. He expressed their metabolic rates in terms of kilocalories per day, a kilocalorie being the unit of energy that weight-watchers and others refer to simply as a calorie. When Kleiber plotted metabolic rate against body mass he obtained a curved graph. When re-plotted with logarithmic data he obtained a straight line with a slope of 3/4, showing that large mammals have relatively lower metabolic rates than small ones. Similar graphs have been obtained for other mammals, birds, mammals and birds mixed, and for ectothermic and poikilothermic animals. If all these animals are plotted on a single graph, we obtain three separate straight lines, each with a slope of 3/4. The three lines correspond to: endotherms (birds and mammals), ectotherms and multicellular poikilotherms (reptiles, amphibians, fishes, and invertebrates), and unicellular poikilotherms (single-celled animals). As noted earlier, the metabolic rate of each group is about ten times higher than the group beneath it.

The fact that such diverse animals have the same exponent provoked much debate on its possible biological significance. Since most of the studies focused on mammals and birds, which are endothermic, the question naturally arose whether the metabolic rate rises more slowly with size because large animals retain heat better. Large mammals have a relatively small surface area through which heat can escape from the body. Small mammals, in con-

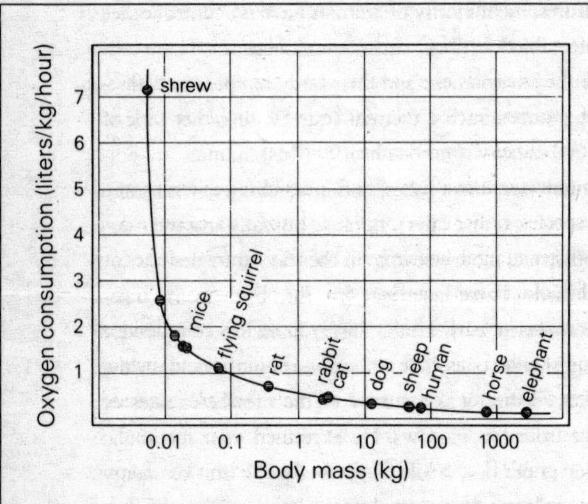

Metabolic rates decrease with increasing body size. Here metabolic rate is measured in oxygen consumption per unit of body mass. For convenience the x-axis is logarithmic (otherwise not all the animals would fit on the graph).

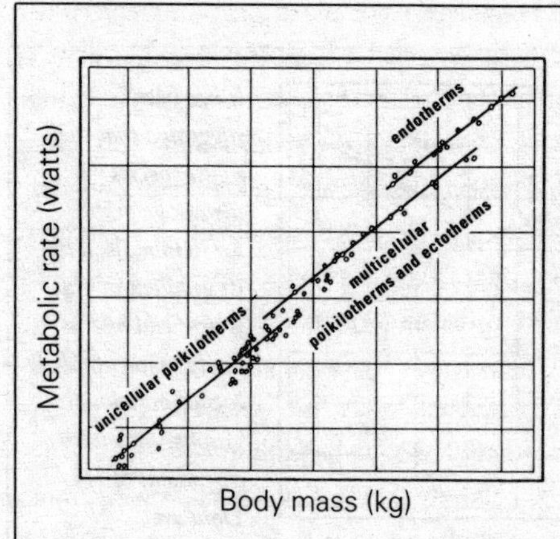

Metabolic rates plotted against body size for endotherms (mammals and birds), multicellular poikilotherms and ectotherms (fishes, amphibians, reptiles, and invertebrates), and unicellular poikilotherms like amoebas and bacteria. Data are logarithmic.

trast, have a relatively huge surface area. They also have the additional problem that they are too small to carry a thick coat: Imagine a mouse with thick fur like a bear's! Some people argue that heat retention explains the relationship between metabolic rate and body mass—small animals need higher metabolic rates to compensate for their greater heat loss, while the reverse is true for large ones. Not so, say others. If this were the case the exponent would have a value of 2/3, not 3/4. Besides, the relationship also applies to poikilotherms, and they have no "interest" in retaining body heat.

The story takes an interesting turn if graphs are plotted for different sized individuals of the same species, rather than for different sized species of mammals. Here, as before, we get straight-line graphs, but this time the slopes are 2/3 and not 3/4. One physiologist has suggested that the multi-species graph, with its slope of 3/4, is really just an artifact made up of lots of individual graphs, each with a slope of 2/3. If this were so, it could vindicate the surface-area argument, except that it cannot account for animals that are not endothermic. A good case has also been made that the numerous points on the multi-species mammalian graph fit onto two separate lines, each with a slope of 2/3, rather than a single line with a slope of 3/4. This would seem to be

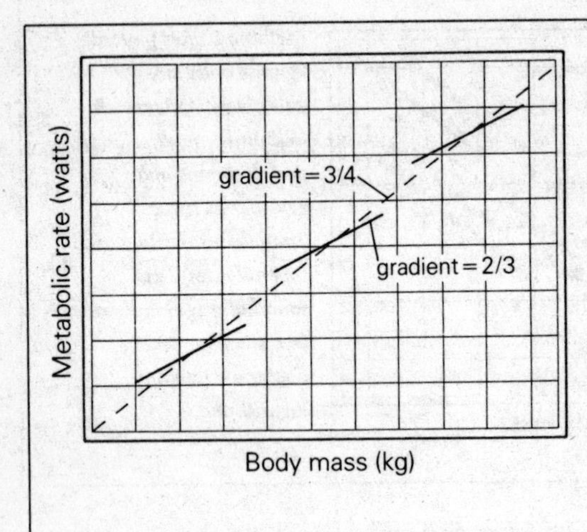

It has been suggested that the multi-species metabolic graph for mammals, with its gradient of 3/4, is artificial and is really made up of lots of individual graphs, each with a gradient of 2/3. Data are logarithmic.

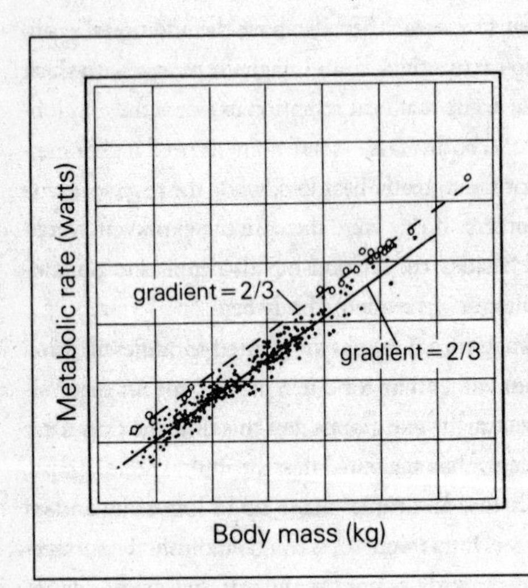

A study based on basal metabolic rates for 391 mammalian species suggests that two lines should be fitted to the data, each line having a gradient of 2/3. Data are logarithmic.

additional support for the surface-area explanation, but it has been argued that the slope of 2/3 is merely a consequence of plotting power (in this case metabolic rate) against mass. These different arguments have their merits, and the matter is far from settled.

Before leaving the subject of metabolism I would like to offer a word of caution about measuring basal metabolic rates. Basal metabolic rates, you recall, are measured when the animal is resting, fasting, and, if it is an endotherm, within its thermal neutral zone; otherwise the ambient temperature has to be at some specified value (often 20° C). These conditions are not difficult to achieve for a hamster or a mouse, but consider the problems of trying to keep an active animal like a tuna or shark at rest when it has to keep swimming to ventilate its gills. Similar problems exist for lots of other animals, including most insects, which seem to be on the go all the time, and for active crustaceans like shrimps and their planktonic relatives the copepods. Reptiles, in contrast, spend most of their time at rest, but their metabolic rates are very dependent on ambient temperatures and other factors like light levels and their previous thermal regimes. Some caution is therefore needed when comparing basal metabolic rates across a broad spectrum of animals.

Body size has some interesting implications for the thermal strategies of non-endothermic animals. One of the most controversial issues in recent years pertains to dinosaurs. Were dinosaurs warm-blooded like mammals and birds, cold-blooded like reptiles, or somewhere in between? Modern reptiles, we have seen, have metabolic rates approximately one-tenth those of comparably sized birds and mammals and have to rely on the sun for most of their heat. Since the sun's heat is absorbed through the skin, small reptiles, by virtue of their larger area-to-volume ratio, heat up more rapidly than larger ones—they also cool down more rapidly. A small lizard of one or two ounces (28–56 grams) needs only ten minutes or so in the sunshine to warm itself up, and it also cools down rapidly when it retreats to the shade. This is handy to know if you want to catch a lizard. Just chase it into the shade and keep it there until it has slowed down sufficiently to be grabbed. Large reptiles, in contrast, take much longer to heat up and cool down. The Galápagos tortoise, one of the largest living land reptiles, has a body mass of about 450 pounds (200 kilograms). Although ambient temperatures fall about 36 F degrees (20 C degrees) at night, their body temperatures fall only 5 F degrees (3 C degrees). By virtue of their large mass, then, their body temperatures remain almost constant night and day, a thermal strategy referred to as *inertial homeothermy*. During a visit to the Galápagos Islands, I had the opportunity to see the effectiveness of this thermal strategy. Our small group had climbed to the top of a 3000-foot (1000-meter) volcano and had camped overnight on its rim. It had been quite hot during the day, but temperatures plummeted during the night and I awoke the following morning feeling decidedly cold. When I emerged from my tent I encountered a tortoise that had spent the night around our camp. He seemed more alert than I was, and when I felt under his armpits with my cold fingertips he felt quite warm. Although large by modern standards, the Galápagos tortoise is minuscule compared with most dinosaurs. As an inescapable consequence of their large body mass, therefore, all but the smallest dinosaurs would have maintained constant and high body temperatures. Indeed, the largest of all, sauropods like *Brachiosaurus* (weighing an estimated 78 tons), probably had trouble getting rid of excess heat as a consequence of their small area-to-volume ratios.

Modern elephants face similar heat dissipation problems. The African elephant, the largest of the two living species, weighs about 6 tons. They have huge ears, with large area-to-volume ratios, which are used as radiators for getting rid of excess heat. These ears are covered by thin skin and are richly supplied with blood vessels; when elephants become particularly hot they flap their ears. Although important sites for shedding heat, elephants' ears account for less than 8 percent of the total heat loss; the remainder is lost through the skin covering the body. Heat loss is facilitated by the high thermal conductance of the skin, high because it lacks fur. What is more, some areas of skin, referred to as thermal windows, have a particularly good blood supply, and this, too, helps in the cooling. Elephants also modify their behavior during hot periods by seeking shade and by wallowing in water when they find any. Sauropods may have had similar strategies for shedding excess body heat.

At the other end of the size range are small warm-blooded animals like hummingbirds, which weigh only a few grams. Their big problem is to retain body heat, since they have such a large area-to-volume ratio. This is not so much of a problem during daytime when their fires are stoked by nectar, but it becomes critical at night. Being small, they have high metabolic rates. But it would be economically unsound to use up valuable fat reserves at night to maintain this high metabolic rate, so they resolve this nightly problem by dropping their body temperature to ambient levels. As their temperature falls they enter into a trance-like state called torpor. Then, shortly before dawn, they begin vigorous shivering, using some fat reserves to fuel their metabolism and elevate their body temperature. One bird that was monitored was found to increase its heart rate from about 150 beats per minute at a body temperature of 70° F (21° C) to over 1000 beats at 97° F (36° C). Once a hummingbird has reached its normal working temperature it takes off and begins feeding again. The downside of sleeping deeply at night is being vulnerable to predators, and not all hummingbirds enter into torpor.

We live beside a golf course that was once an apple orchard, and each fall we have to contend with a rain of maggoty apples. And with the apples come the wasps. When I have to pick up an apple with a wasp feeding on it I first nudge it with my foot and the wasp usually flies off. But if we have a cold snap at this time of the year, apple gathering becomes a hazardous affair. When temperatures fall below the operating temperature of the wasps' flight muscles, they are unable to take off.

Lots of other flying insects are grounded by the cold, too, but some of them, including several species of bees, moths, and beetles, can overcome the problem by heating their flight muscles prior to takeoff. They do this by rapidly contracting the muscles that elevate and depress the wings, which causes a shivering movement. The cells of their flight muscles are rich in mitochondria and therefore generate large quantities of heat when they are active. Flight muscles are located in the thorax—the body segment next to the head bearing the wings and legs. The thorax is well insulated with a hairy covering that reduces the heat loss to the environment. Once flight muscles are hot enough the insect can take off, and the heat generated during flight keeps its thorax warm. Much of our knowledge of these heterothermous insects has been derived from laboratory experiments in which tethered bees and moths have been induced to go through their warm-up procedure. Imagine the ingenuity required to measure a bee's thoracic temperature with tiny electrical thermometers implanted into its thorax! Since pre-flight warmups require energy, repeated laboratory trials fatigued the bees, which meant longer warm-up times and lower thoracic temperatures. The remedy: feed them with sugar solution. It is not unusual to see grounded bees stumbling around in the grass during the summer months, their fuel reserves spent. If fed sugar solution they soon recover and fly off. I have never refueled a bee myself, but McGill University ecologist Robert Peters has, and he tells me it works like a charm.

Large heterothermous insects have a thermal advantage over smaller ones because they retain heat better. This can be seen rather nicely in a bee called *Anthophora plumipes*. Large individuals of the species have higher thoracic temperatures at the end of their pre-flight warmup and during flight than do small ones. The female of the species is consistently larger than the male, which probably explains why females can fly on colder days when males cannot.

Water has a much higher thermal conductance than air, which is why a frozen chicken thaws much faster if left in water than on the kitchen table. This accounts for why we cool off so rapidly when we take a summer dip. Indeed, long swims drain so much heat energy from the body that you feel quite cold afterward, even at the height of summer. During a ninety-minute swim in July, when it was 80° F (27° C) in the shade and the water was a balmy 75° F (24° C), I felt frozen when I got out of the water; my body temperature had dropped almost 2 F degrees (1 C degree). Aquatic endotherms—seals, dolphins, whales, polar bears, and the like—have therefore adapted to conserve heat. Large size is a major factor, and most marine mammals have large bodies, and hence small area-to-volume ratios.

Body insulation is clearly important in heat conservation. Seals have waterproof fur that is oiled by sebaceous glands in the skin. This fur is especially dense in fur seals, and creates an effective insulating layer. Like a scuba diver's rubber suit, though, the fur becomes compressed with depth, and therefore less effective; to compensate the seals have a thick layer of blubber beneath the skin. Polar bears also have a thick coat, which sheds water readily when the bear shakes itself off after swimming. Like seals, they also have a thick layer of subcutaneous fat. From the perspective of absorbing radiant heat from the sun, their white fur is the worst possible color and black would be best (our black cat's fur feels warm when she lies in the sun). However, camouflage is important to them for sneaking up on their prey. As a compromise, their skin is black, and readily absorbs the radiant heat channeled down by their fur. Adult polar bears, which often weigh over 1200 pounds (450 kilograms), have no problem staying warm in cold Arctic seas, but their cubs have considerably larger area-to-volume ratios, and are vulnerable to cold water.

Cetaceans (whales and dolphins) rely entirely on a thick layer of blubber for insulation. Although heat retention is their major thermal problem, they must also be able to lose excess heat to maintain thermal equilibrium when, for example, they are swimming in warmer waters. Most excess heat is shed through their fins, which are thinly insulated and have a rich blood supply, just like elephants' ears. Their bodies are continuously washed with fresh seawater, a major factor in their thermal budget, and overheating results if this is interrupted, as when whales become stranded on the shore. When cetaceans are being transported by marine aquariums, precautions against overheating must be taken, particularly with larger species like killer whales that need to be constantly hosed down with water.

All of the aquatic mammals mentioned so far have large bodies and therefore small area-to-volume ratios to aid heat conservation. But the muskrat, which faces North American winters without hibernating, is only 16 to 25 inches (41 to 64 centimeters) long and weighs 3 pounds (1.4 kilograms). Like seals they have a thick oily fur, but since they are so small they cannot conserve heat so well. They resolve the problem by elevating their metabolic rate and hence their body temperature before entering the water to feed. They also limit the time they spend in cold water. When a muskrat swims beneath the ice its body temperature drops a few degrees, but this soon returns to normal levels when it leaves the water.

Large fishes, with small area-to-volume ratios, have the potential to retain the body heat generated by their active muscles, but this is compromised by their piscine respiratory system. Fishes respire through gills, whose feathery structure maximizes the area-to-volume ratio so the oxygen-depleted blood flowing through the gills can absorb the greatest amount of oxygen from the surrounding water. Respiratory exchange is also enhanced by close contact between the blood within the gills and the water outside of them. A consequence of these features is that the blood cools down to the same temperature as the water. All of the heat generated by active swimming muscles is therefore lost when the blood passes through the gills.

But in some fishes, notably the tuna and its relatives, a modification of the circulatory system has evolved. Heat exchange occurs in a network of fine blood vessels, a network described as a countercurrent heat exchanger. Heat from the warm part of the network passes into the cold network so that the blood gives up its heat to the body before the heat is lost through the gills. It works like two car radiators—one carrying warm fluid, the other carrying cold fluid—that have been squashed together. Countercurrent heat exchangers are found in many other locations too, including whales' flippers and birds' feet. They enable a whale to restrict heat loss from its flippers when it needs to conserve heat, and explain how penguins can stand on the ice without losing very much heat.

Tunas, like their relatives the mackerel, sailfish, and swordfish, are active fishes that swim all the time. One school of tunas that was monitored covered a distance of 266 miles (428 kilometers) in a single day, achieving an average speed of 11 miles per hour (18 kilometers per hour). Their body heat is generated mostly by active swimming muscles, whose two main types are red and white in color. (I will discuss an intermediate type in the next chapter.) Red muscle cells are rich in mitochondria and are therefore metabolically highly active, generating large quantities of heat. These muscle cells are used for continuous cruising, whereas the white muscle cells are primarily for quick bursts of speed. A tuna's red muscle cells are concentrated in blocks of red-brown muscle that lie deep within the rest of the body muscle; in other fishes, like mackerels and marlins, the red muscle blocks are placed more peripherally. Next time you eat a mackerel or a herring look out for its red muscle—it is a dark colored strip running along the side of the body, immediately beneath the skin. Tunas are large fishes, reaching lengths of up to 10 feet (3 meters) and weighing as much as 1400 pounds (650 kilograms). The largest individuals maintain the highest body temperatures—about 95° F (35° C)—which seems all the more remarkable when you consider the cold depths where they swim. Even small skipjack tunas, weighing only a few pounds, can achieve body temperatures of 37° C in waters of 24° C, but only when active.

Swordfish are also large, reaching more than 12 feet (4 meters) long and weighing more than 1000 pounds (450 kilograms), and are noted for their long excursions from the surface, where summer temperatures reach 80° F (27° C), to depths of 500 feet (150 meters) and more, where temperatures drop to about 46° F (8° C). Unlike tunas they do not maintain warm bodies, but they have a brain heater that maintains the temperature of the brain and eyes. This enables the visual and cerebral activities of these active predators to be kept at optimum levels in cold waters. The brain heater lies deep in the head between the eyes; it is actually a modified eye muscle, one of several pairs used for moving the eyeball about in its socket. The muscle's cells are rich in mitochondria and most of them have lost their contractile apparatus and cannot contract. It should come as no surprise that it was the eye muscles that became modified into a brain heater because they are already metabolically very active. They are also in a prime location for their new role because they lie so close to the brain. The swordfish's relatives—marlins, sailfish, and spearfish—have fewer of these specialized cells in the brain heater. These fishes spend less time at great depths, and their brain heaters presumably generate less heat.

Our look at endothermic organisms has been mostly concerned with adults, but their offspring face some interesting problems, primarily because of their small size. All young endotherms pass through this transitional phase during their development.

When a mammalian fetus is inside the uterus it receives what it needs through the mother's blood, functioning as if it were part of her body. Like any other part of the maternal body, its metabolic rate is at a level commensurate with that of the larger mother, rather than that of the smaller offspring. This is true for our own species; hence when a baby is born its metabolic rate is considerably lower than it should be for a mammal of that particular body mass. Within about thirty-six hours of birth, though, the infant's metabolic rate rises to a level appropriate for its size. Other placental mammals presumably go through a similar metabolic transition after their birth. An important source of heat production in young mammals is provided by the respiration, or "burning," of a special fat. Called brown fat because of its color, it differs from ordinary white fat in that it is very easily respired. (The fact that white fat is so difficult to respire explains why we have such problems getting rid of it when we are trying to lose weight.)

Birds, in contrast to most mammals, lay eggs that must be kept warm once development starts—otherwise the embryos die. Except in large species like the ostrich, birds lay small eggs with thin shells, and their large area-to-volume ratios cause them to lose heat readily. Parent birds, as we all know, sit on the eggs to keep them warm. Only brief absences are tolerated, especially for the smallest eggs like those of small finches. When the chicks hatch out they are either small, naked, and helpless, like those of sparrows and thrushes and most other birds found in the garden, or they are large, fluffy, and capable, like hatchling chickens, ducks, and gulls. The helpless chicks, called *altricial* (Latin, *altrix*, "a nurse"), cannot keep themselves warm and rely on extensive parental care, not only for heat but for food. The larger chicks, called *precocial* (Latin, *praecox*, "early ripening"), can maintain their own body temperature, and usually fend for themselves. Experiments have shown that eggs that have been cooled by a few degrees respond by generating additional heat. So we know that precocial chicks can generate body heat for a few days before they hatch.

Keeping eggs warm consumes time and energy. Some birds have overcome the problem by laying their eggs in mounds of rotting vegetation. The heat generated by the decomposition incubates their eggs. Some reptiles do the same, but the megapodes, a group of Australian birds related to the chicken, have perfected the technique. Mated birds construct a large mound of rotting vegetation, about 4 feet (1.2 meters) high and 16 feet (5 meters) in diameter, a mass of several tons. This is maintained throughout the breeding season, which lasts from spring to fall. Decomposition raises the temperature of the mound, as with a compost heap, and when it reaches about 90° F (33° C) the female commences laying. The birds monitor the mound's temperature; when it is too cold they add more vegetation, and when it is too hot they remove material, piling on sand in midsummer to prevent overheating. In this way the birds can maintain temperatures within a few degrees. The megapodes also regulate the mound's water content, keeping it at a level that minimizes heat loss through conduction and regulates the decomposition rate; the result is that they have less refueling to do. They apparently regulate the water content by modifying the shape of the mound. When they need to add water they excavate a large depression in the top to trap rainwater; they fill this in with vegetation when they want to keep the water out.* The mound's heat output is about 100 watts—more than twenty times the power that the birds could generate themselves. Such elaborate mounds enable the megapodes to incubate far more eggs than would be possible in a conventional nest.

We have seen that size has a profound effect on the thermal strategies of animals, especially those that generate their own heat. Size, as we know from our own experience, is also highly correlated with strength, which is the subject of the next chapter.

*The information about how megapodes control the water content of their mounds was generously supplied by Roger Seymour of the University of Adelaide. The idea was first suggested by Australian naturalist David Fleay in 1937 and has since been confirmed by Roger Seymour's own observations.

PUMPING IRON

When the circus came to town it was the entertainment highlight of the year—animals, performers, sideshows—all the fun of the fair. Not all circuses were grand enough to have a big top with lions and tigers and elephants, but in bygone days these shortcomings were made up for by equestrians, acrobats, jugglers, strongmen, and every sideshow attraction beneath the sun. Back then if you paid your penny and stepped inside the flea circus, you would be treated to a procession of miniature carriages and Roman chariots, all purportedly drawn by fleas. The showman would expound on the herculean strength of fleas that could pull vehicles many times their own weight, but it is likely that the only fleas in the tent were those on himself and his patrons!

Regardless of the veracity of flea circuses, the remarkable strength of insects is not in doubt, as most of us have witnessed. Consider the ease with which ants carry objects heavier than themselves—or how dung beetles, dwarfed beside huge balls of manure, spend their lives rolling them across the African countryside. To put these feats of strength into perspective, how heavy a weight can you lift? I weigh 165 pounds (75 kilograms) and can lift a 100-pound (45-kilogram) bag of plaster without much trouble, but I would not want to carry it very far. This is a puny effort compared with an ant's, but how does it compare with that of a really large animal like an elephant? Elephants are certainly strong, for they can push over large trees with ease. The Indian species often serves as a beast of burden for hauling timber. When an elephant picks up a 12-foot (3.7-meter) tree trunk 2 feet (0.6 meter) in diameter and weighing 2000 pounds (910 kilograms), it seems to be an incredible feat of strength, but is it? Indian elephants weigh about 9000 pounds (4000 kilograms), so a large tree trunk represents about 22 percent of its body mass—the equivalent of my picking up an object weighing 36 pounds (16 kilograms). Not very impressive. Elephants, then, are quite weak, much weaker than humans on a weight-for-weight basis, and positively feeble compared with ants. So why are small animals like ants so strong while large animals like elephants are seemingly so weak? To answer this we need to define the term "strength."

One convenient way of measuring an animal's strength is to see how much weight it can lift. Suppose that you have started working out at a gym and want to assess your strength. Your upper-body strength could be assessed by a bench press, where you lie on your back and push up a set of weights. You are lying on your back, gripping the weight's cross-bar with both hands, and are beginning to push. Perhaps you've loaded too many weights on the bar because it feels as if it's bolted down! You are now pushing with all your might and the weights begin to lift. At the point when the weights move, the force exerted by your arm and shoulder muscles is equal to the force that gravity exerts on the weights. Force is the product of mass and acceleration, specifically the acceleration due to gravity. Gravity is the mutual attraction between bodies that makes them accelerate toward each other, and is measured in meters per second per second (m/s^2).

If your car accelerates at 1 meter per second per second, it means that for every second you hold the gas pedal down, your speed increases by 1 meter a second. At the start your velocity is zero, but at the end of the first second you are traveling at 1 meter per second (about 2 miles per hour). At the end of the next second you are traveling at 2 meters per second (about 4 miles per hour), and at the end of 10 seconds you're traveling at ten meters per second. This is only about 20 miles per hour; so an acceleration of 1 meter per second per second, written in shorthand as 1 m/s^2, is not much of an acceleration for a car. The acceleration due to gravity is 9.81 m/s^2.

To see how forces are measured numerically, suppose that you managed to lift 60 kilograms (132 pounds). The force of gravity acting on the set of weights you have just raised is obtained by multiplying their mass (60 kilograms) by the acceleration due to gravity (9.81 m/s^2), and the resulting units are expressed in *newtons* (named after Sir Isaac Newton and abbreviated N). This force is therefore $60 \times 9.81 = 588.6$ newtons and is equal to the force exerted by your muscles. (The muscles actually exert much larger forces than this because they work at a mechanical disadvantage.)

Muscles can only pull (that is, exert a tensile force); they cannot push to exert a compressive force, despite what you may have thought from the weight-lifting exercise. Consequently every moveable joint in the skeleton has to have at least two muscles to work it: one to pull the joint open and one to pull it closed. The elbow joint, for example, has two main muscles: biceps close the joint (flexion), causing the hand to move toward the shoulder, and triceps open the joint (extension), moving the hand away from the shoulder. So when you pushed that heavy set of weights the relevant sets of arm and shoulder muscles pulled against your skeleton to raise your arms. You could tell which muscles were generating force because they were the ones that were bulging. They were bulging because they were getting shorter.

If you have not done any weight training before you may be feeling a bit disappointed with your performance. And to add insult to injury a guy on the next bench is bench-pressing a 200-pound (90-kilogram) bar as though it weighed nothing at all! One look at his physique provides obvious proof that large muscles correlate with strength. To understand why requires a knowledge of the anatomy of muscles and how they work.

In its simplest form a muscle is an elongated structure composed of numerous bundles (called *fascicles*) of muscle cells that run from one end to the other. This structure is enclosed in a tough sheath that is continuous at either end with a cord-like attachment, a tendon, which anchors the muscle to the skeleton. Muscles contract by about one-quarter of their length, and since their volume remains the same, it follows that they bulge when they contract. Individual muscle cells have diameters of only 10 to 100 micrometers. (A micrometer, also called a micron, is one thousandth of a millimeter—a human red blood cell is about 7 micrometers in diameter.) But these cells are quite long, sometimes as long as the entire muscle, which is why they are often referred to as muscle fibers. Packed together tightly like drinking straws in a box are fibrils, which are fine structures with even finer threads, called filaments, inside them. Two sorts of filaments, made of protein molecules of actin or myosin, interdigitate with one another. When a muscle cell is extended the overlap between actin and myosin filaments is minimal, and the muscle cell is at its longest. When a muscle cell shortens the filaments pull past each other and overlap more, shortening the cell's length. The contractile mechanism works something like a ratchet, with cross-bridges being made and broken between actin and myosin filaments.

A muscle's force of contraction is proportional to how many actin and myosin filaments overlap, which is, in turn, proportional to the cross-sectional area of the muscle cell. Hence, the force generated by an entire muscle is proportional to its cross-sectional area, which helps explain why the guy with the big muscles on the next bench can lift heavier weights. His weight training has increased the size of his muscles not by adding more muscle cells, which remain essentially unchanged throughout adult life, but by enlarging the size of each cell.

How does the nervous system regulate muscle action? Nerve impulses stimulate muscle cells to contract, and when the cells receive these electrical im-

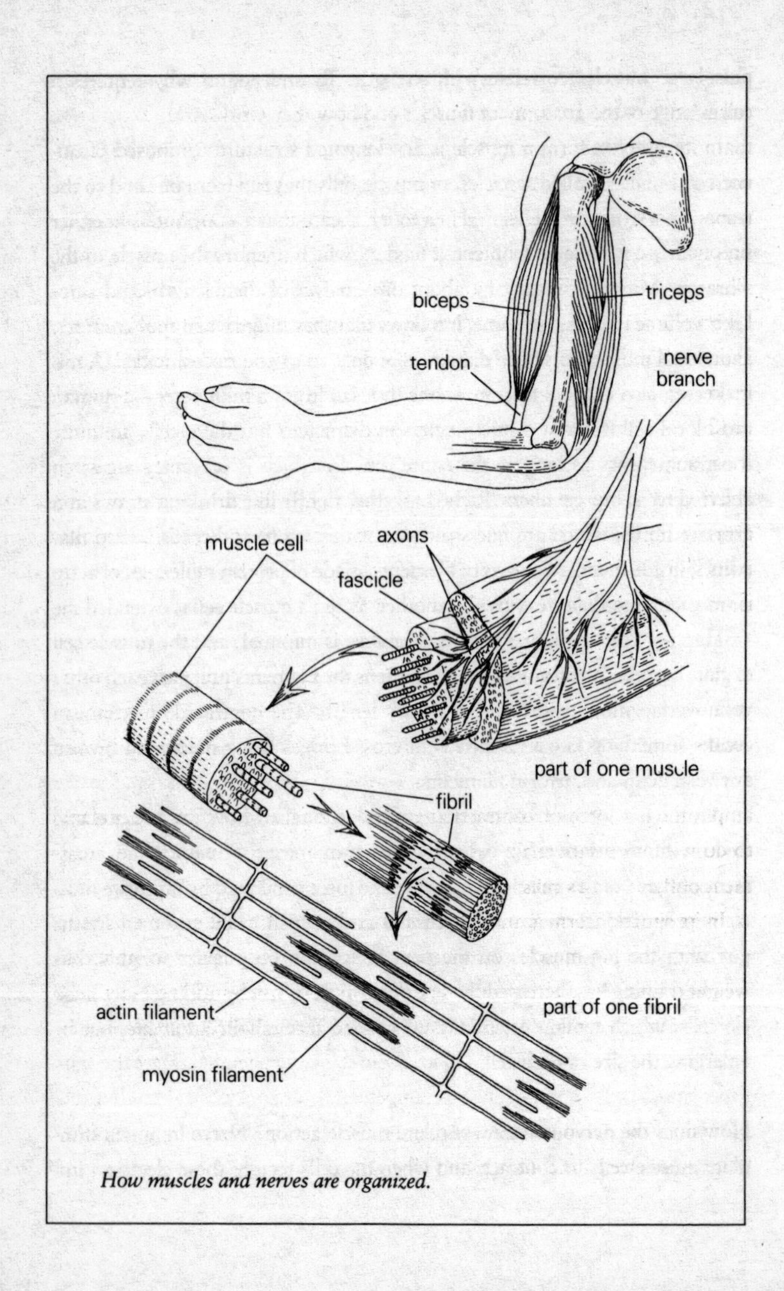

How muscles and nerves are organized.

pulses they contract maximally. There are no partial responses; the muscle cell either contracts fully or not at all. But we know from everyday experience that muscles do not contract with full force every time. If they did you could not scratch the end of your nose without knocking yourself out! The graded response of whole muscles is achieved by how they are innervated (that is, supplied and stimulated by nerves). Each muscle is supplied by a nerve, or by a branch of a nerve, made up of bundles of individual strands, called *axons*. Like wires inside a telephone cable, axons transmit electrical impulses. Each axon ends in several branches, like the roots of a plant, and each branch makes contact with a single muscle cell. A particular axon, with the several muscle cells it supplies, is called a *motor unit*. When a certain nerve cell fires it contracts all the muscle cells in its motor unit. When you raise your arms above your head, you are contracting most of the skeletal muscles in your arms and shoulders, but you are stimulating only a small percentage of motor units in those muscles. But when you bench-press a set of weights, your effort is maximal and fires all the motor units in the relevant muscles.

The fact that strength increases with the cross-sectional area of muscles explains why small animals are relatively stronger than large ones: area-to-volume ratios are larger for smaller bodies. This relationship adequately accounts for differences in relative strength between ants and elephants. Like any generalization, though, one can always find exceptions. Take, for example, the prodigious crushing strength of a lobster's claws. This has nothing to do with the relationship between area and volume but with the arrangement of the muscle's cells.

In its simplest form a muscle's cells extend from one end of the muscle to the other; so the muscle's contraction force is in proportion to its cross-sectional area. An alternative arrangement occurs when much shorter cells run obliquely across the muscle. This arrangement is called *pennate* (sometimes written *pinnate*), because of its resemblance to a feather. Here the cells either attach on the opposite side (unipennate), or to a central tendon (bipennate). Like a non-pennate muscle the muscle cells shorten by about one-fourth of their original length, but they are considerably shorter and so the

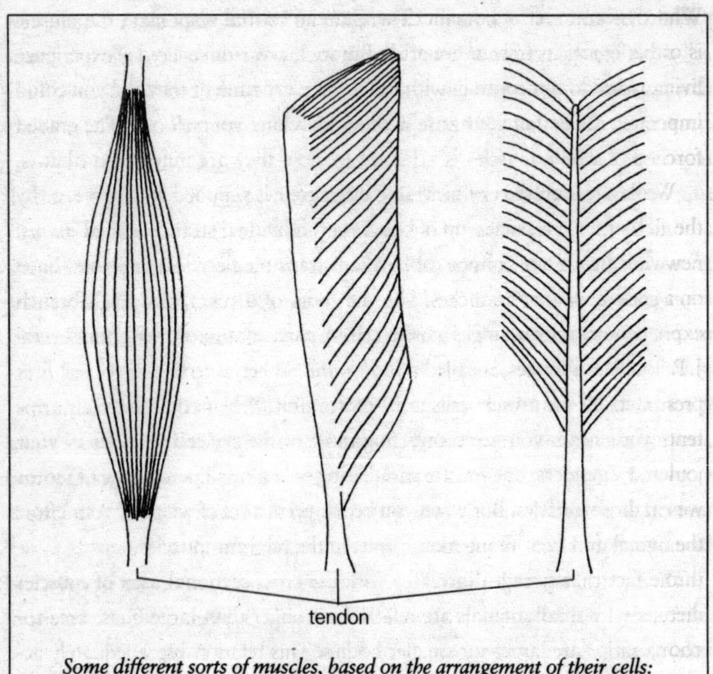

Some different sorts of muscles, based on the arrangement of their cells: non-pennate (left)*, unipennate* (middle)*, and bipennate* (right)*. The individual muscle cells are too small to be seen and the striations are due to the bundles* (fascicles) *of cells.*

whole muscle shortens only a little. Yet since pennate muscles have far more muscle cells than do non-pennate ones, they contract more forcefully.

The muscle that closes the lobster's claw is of the bipennate variety that uses more cells and contracts with greater force. Indeed, the force is so great that the live lobsters you see at fish bars usually have their claws bound to prevent them from inflicting severe wounds on careless fingers or each other. The next time you crack open a lobster claw, look at the fiber arrangement of the large muscle inside. The thin white plate running down the muscle's center is the central tendon that attaches to the bundles of muscle cells.

While the question of how much weight can be lifted in a static bench press is of biological interest, the world of motion remains to be explored. In the living world of swooping hawks and leaping fishes, dynamics is usually more important than statics; that is, work and power are more important than force.

Work is the product of force and distance. If I want to pick up a book off the floor that has a mass of 1 kilogram, I have to exert a force of 1×9.81 newtons just to free it from the carpet. If I lift the book 2 meters and put it on a shelf, I will have performed $1 \times 9.81 \times 2 = 19.62$ units of work, which are expressed in terms of *joules* in metric units (named after the English physicist J. P. Joule and abbreviated J). Energy is the potential to do work and is expressed in the same units as work. That book sitting on the shelf has the potential to do work—19.62 joules worth of work. If it fell to the floor those joules of energy would be released. We often express the energy of the food we eat in *calories*, and 4.2 joules are equivalent to one calorie. A calorie is the amount of heat required to raise the temperature of one gram of water through one centigrade degree. The calories referred to in cookery books and diet sheets are actually kilocalories or large calories, being the equivalent of 1000 calories.

We know that if we had two muscles of the same volume, one short and fat and the other long and slim, the first would generate a large force over a short distance while the other would generate a small force over a long distance. The first muscle exerts more force than the second, and the second causes a greater movement. In terms of work performed, though, they are equal. This means that although small animals can generate more force relative to body mass than large ones—by virtue of their muscles' larger area-to-volume ratios—they perform the same amount of work relative to body mass (if we assume that the muscle mass and body mass are proportionately the same for both animals).

Power is the rate of doing work and since small animals generally have faster muscles, they produce more power per unit of body mass than large animals. So if a mouse and an elephant had a race, the mouse's muscles would generate more power. Power is measured in watts; a watt is the amount of power required to perform 1 joule of work per second. If two people who weigh the same set off down the street, one walking and the other running, the runner's leg muscles would generate more power than the walker's. These concerns with muscles and scale bring us to the skeleton and how stresses on bones vary according to body size.

If two animals were identical except for size, and one was twice as large as the other, the bones of the large one would experience twice as much stress (force per unit area) as those of the other. The difference in bone strength from animal to animal is small compared with the variations in the stresses on the bones of different-sized animals. Like any other material, however, bone has limits to its strength; if stressed enough it will break. For this reason some strategy is required to avoid exceeding the breaking point. The most obvious solution is for the weight-bearing bones—primarily the leg bones—to become thicker as the body gets larger.

Compare the femur (thigh bone) of a deer with that of a dog and note how much thicker the deer femur is. The femur of a hippopotamus is chunkier still, and the same trend is seen in other bones and for other animals, too. The idea that large animals have more robust bones than smaller ones makes good sense and was once widely accepted. But when investigators began looking at a wider range of animals, they found all manner of exceptions. Elephants, for example, have remarkably slender femora. When data for numerous mammals are assessed, one finds that the thicker the leg bones, the longer their length.

As large mammals do not have more robust bones than small ones, how do they prevent their bones from breaking under the burden of their massive bodies? The first thing to point out is that bones do not usually break when an animal is standing still: they invariably break when an animal is in motion because that is when the stresses on the bones are highest. Large animals therefore take care not to put undue stress on their skeletons during locomotion. A good illustration of this is the careful and deliberate way that elephants move. They do not gallop or jump fences, and although they can run quite fast (up to about 20 miles or 35 kilometers per hour), they do so with a straight-legged gait that looks odd to us but that minimizes the stresses on

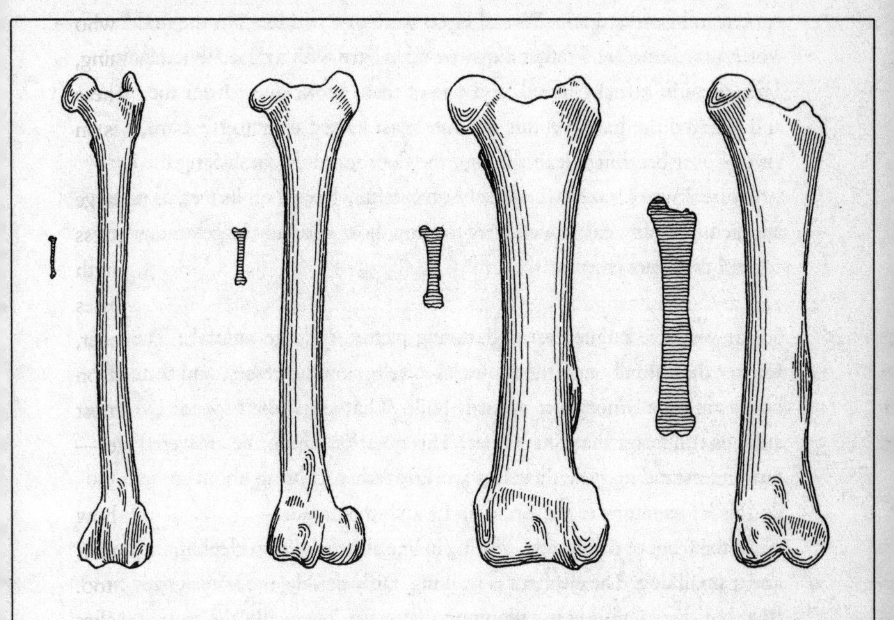

Left to right: *Thigh bones (femora) of the dog, deer, hippopotamus, and African elephant. Large animals are often thought to have relatively thicker bones than smaller ones. While this may be true for selected groups, as depicted in the first three femora, it does not hold for a wide range of groups. The relative lengths of the bones are shown in silhouette.*

their leg bones. Forces that act straight along a bone produce much less stress than do forces that bend it. By keeping their leg bones more nearly vertical than smaller mammals do, large mammals keep the stresses in their bones about the same as those of smaller animals during strenuous activity. The safety factor of bones—the breaking stress divided by the peak activity stress—is therefore independent of size. Safety factors have been determined experimentally for mammals ranging in size from squirrels to horses and range from about 2 to 4.

I am reminded of an incident several years ago that happened with an

American bison at a zoo. Several bison were in a paddock, in the middle of which was a small and rather flimsy wooden barn with a raised wooden floor. They were in a frisky mood, and one of them broke away from the others and charged the barn. As this one-ton beast leaped up into the barn, I had visions of it breaking clean through the floorboards, demolishing the entire structure. I was amazed when the bison carefully landed on its feet, as nimbly and gently as any cat, which goes to show how carefully large animals can control their movements.

So far we have painted an unflattering picture of large animals: They are weaker than small ones; their muscles develop lower stresses; and their limb bones are usually not more robustly built. What about their speed? Do large animals run faster than small ones? This question cannot be answered without understanding how their legs work together to bring about motion. To do this let us return to the circus and watch the parade.

At the front of the parade, moving in line abreast, are an elephant, a horse, and a small dog. The elephant is walking, fairly briskly, the horse gently trotting, but the small dog is galloping to keep up. The walk, the trot, and the gallop are three distinct patterns of locomotion, called *gaits*, that four-legged animals (quadrupeds) use to move at higher ranges of speed. The three gaits are analogous to three forward gears of a car. At slow speeds quadrupeds walk, accelerating by increasing the number of strides taken per minute (a stride is a complete cycle of one leg, from when the foot strikes the ground until the same foot strikes the ground again). But a point is eventually reached when a continued acceleration requires a change in gait from the walk to the trot. The animal "changes gear," but the car analogy is not perfect because revolutions do not drop as with a car. Stride frequency continues to increase throughout the trot, but not so rapidly as it did for the walk.

What differentiates the walk from the trot is the sequence in which the legs are moved. Think of a horse and rider. When the horse is walking it has two or three hooves on the ground at any one time and the equestrian has a smooth ride. As soon as the horse begins to trot, no more than two hooves are on the ground at any one time, and sometimes none are (except during

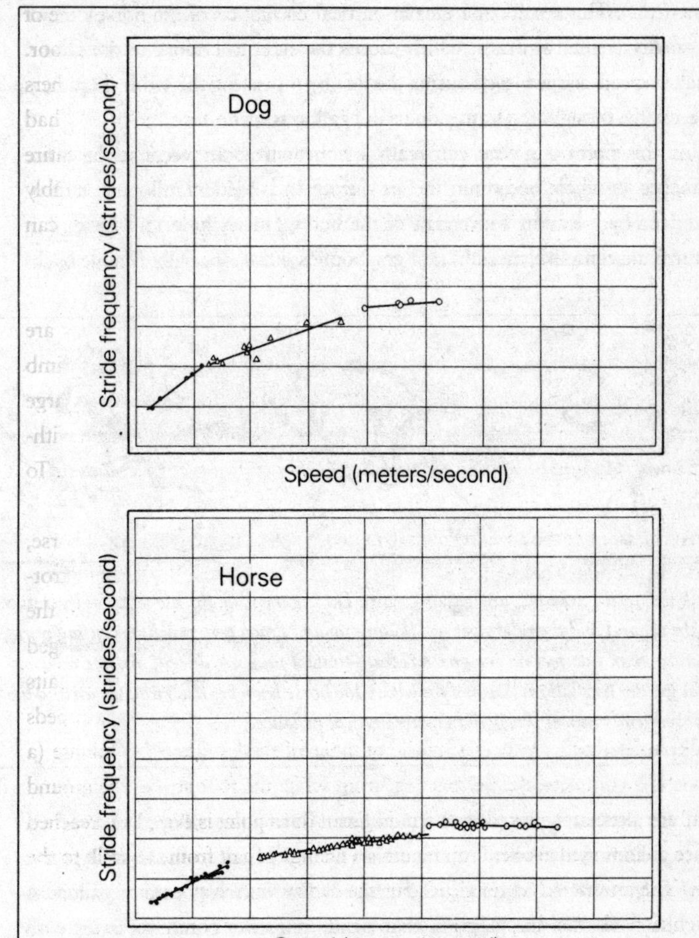

How stride frequencies and speeds change in the walk (dots), *trot* (triangles) *and gallop* (circles) *for the dog* (upper graph) *and horse* (lower graph). *Notice that higher speeds are achieved in the walk and trot by increasing the stride frequency. Higher speeds in the gallop result from increasing the stride length.*

slow trots). This results in a greater vertical oscillation of the horse's center of gravity (center of mass), which causes the rider to bounce in the saddle. Higher speeds are achieved during the trot by increasing the stride frequency. Eventually, though, the transition to the gallop is made.

At this point a novice can really be thrown about because the horse launches its whole body into the air during each stride. Galloping is characterized by a leaping movement of the body. Horses have stiff backs, but some mammals, like cheetahs and greyhounds, have especially flexible backs

The walk (left), *trot* (middle), *and gallop* (right). *During the walk the horse has at least two feet on the ground and the rider does not bounce up and down very much. A trotting horse has no more than two feet on the ground, their contact time is reduced, and the rider experiences greater oscillations. During the gallop the horse launches itself into the air and the rider consequently moves through a greater vertical distance.*

that are alternately flexed then straightened during the gallop, like a spring. Since galloping involves a leaping action from the back to the front legs, it is obviously restricted to quadrupeds; bipeds like ourselves cannot gallop. A singular feature of the gallop is that stride frequency barely increases with speed; higher speeds are achieved by lengthening the stride—taking longer leaps in the air. A graph of stride frequency plotted against speed therefore depicts the gallop as a straight line with a very slight slope. When an animal changes from a trot to a gallop it has reached its capacity for making rapid body movements and is essentially running at its maximum sustained stride frequency.

In the circus parade the elephant, the largest of the three, is walking, the horse is trotting, while the dog, the smallest, gallops to keep up. This is because the speed at which animals change from one gait to another increases with increasing body size, which is true for a wide variety of animals, from mice and squirrels to gazelles and waterbuck. The speed attained at each gait also varies with body size. Therefore if the dog, the horse, and the elephant were walking along the street at a comfortable pace, the elephant would be way ahead of the horse and the dog would be far behind. The same would also hold true for the trot and for the gallop, except that elephants cannot gallop, a restriction imposed by their large size. Stride frequency, however, decreases with increasing body size. A walking dog therefore takes more strides per minute than a walking horse, which in turn takes more strides per minute than a walking elephant. Smaller animals, then, are relatively slower and take more strides than do larger ones for each gait.

Quadrupeds' three gaits cover three speed ranges: they walk at low speeds, trot at moderate speeds, and gallop at high speeds. The transition from trotting to galloping essentially marks the upward limit of stride frequency, and could be regarded as a physiologically similar speed for animals of different sizes. For example, a raccoon and a zebra running at the transition point between a trot and a gallop could be considered to be moving at equivalent speeds, although in terms of miles per hour the raccoon is traveling much more slowly. The conventional wisdom is that the transition from one gait to the next is a strategy for reducing energy costs. This is certainly one of the major rationales for changing to a higher gear in a car, and fuel economy would be seriously compromised if you kept your car in a lower gear. However, Claire Farley and Richard Taylor proposed that the switch in gaits is to reduce the stresses on the skeleton. Their experiments with horses running on a motorized treadmill showed that peak stresses increased with trotting speed, but dropped suddenly when the switch was made to the gallop. Significantly, the switch did not coincide with a reduction in transport costs because the horses started galloping at speeds where it would have been more economical to continue trotting. It occurred, instead, when stresses on the skeleton reached a critical level. Other researchers have argued that the change in gait need not necessarily be linked in a cause-and-effect relationship. They also question how changes in bone stress could be detected by the body and used to trigger gait changes. Clearly, more research is needed to resolve the problem.

Larger animals take larger but fewer strides; conversely, small animals make up for their short stride lengths by taking more of them. This relationship between size and stride has led some biologists to argue that small animals should be able to run just as fast as large ones, and this argument can be supported on theoretical grounds which we will not go into. The idea that speed is independent of size is counter-intuitive and contrary to our everyday experience, but there is experimental evidence for this. In one study investigators clocked running speeds of ten species of African ungulates (hoofed animals) by chasing them in a car. The animals ranged in size from a 44-pound (20-kilogram) gazelle to a giraffe weighing fifty times as much, but they all had about the same top speeds. Regardless of the large differences in size, this is a small sample size, restricted to only one group of animals. Not surprisingly, a different picture emerged from a study of over one hundred species of mammals. This more comprehensive study showed that top running speeds do indeed increase with body size. But the largest mammals are not the fastest; speeds peak at body masses of about 260 pounds (120 kilograms). Some caution is needed when considering running speeds, though, because one encounters many difficulties when trying to record them. Even if you can persuade an animal to run alongside your truck while you clock its speed, the chances of its running in a straight line are remote, which clearly affects the accuracy of the measurement. Furthermore, although the animal may be anxious to escape the vehicle and its annoying occupants, there is no guarantee that it is running at its top speed. Even so, these records of running speeds are the best we have, and since all such experiments suffer from similar measurement errors, they are likely to provide a good indication of the relative running speeds of a wide variety of mammals. They show that, up to a point, speed increases with body size.

An interesting point to emerge from these data is how poorly humans perform. A world-class sprinter is capable of reaching only about 22 miles per hour (36 kilometers per hour), which is on a par with the African elephant, a species not noted for its fleetness of foot! We cannot attribute our poor showing to our being two-legged because other bipeds, like kangaroos and emus which are roughly our size, can run much faster. And our modest 22-miles-per-hour sprint can be kept up only for short distances. World-class marathon runners manage about 13 miles per hour (20 kilometers per hour).

Training for a one-hundred-meter sprint and for the twenty-six-mile marathon requires entirely different strategies, as reflected in the contrasting physiques of the two runners. Sprinters are usually large and muscular, like bodybuilders, whereas long-distance runners are usually small and slightly built. The physiological principles underlying these marked size differences warrant closer inspection.

I jog on a regular basis and can run for several miles at a reasonable pace without getting winded. When I am running at a comfortable pace my muscles are getting oxygen fast enough to supply their needs. Under these conditions the muscle cells are carrying out their metabolism *aerobically*, meaning "with air." During aerobic metabolism the carbohydrate fuel being supplied to the muscles breaks down into carbon dioxide and water. If I started running flat-out I could not supply enough oxygen to the muscle cells for them to continue metabolizing aerobically; most of them would switch over to *anaerobic* metabolism, meaning "without air." During anaerobic metabolism the carbohydrate fuel is incompletely broken down, with the result that lactic acid accumulates in the muscles, making them feel stiff, and eventually stopping their action altogether. This anaerobic process, called *glycolysis*, is far less efficient than aerobic metabolism, and produces less than 10 percent as much energy. Once the anaerobic activity is over the lactic acid has to be broken down with oxygen. This explains why I continue puffing and panting after I finish sprinting; I have to use the extra oxygen to break down the lactic acid. Incidentally, it is mostly the residual lactic acid in muscles that causes stiffness after a run. A massage can reduce the stiffness by coaxing lactic acid out of the muscle and into the bloodstream.

You may recall from chapter 2 that tunas have red and white muscle cells and that I mentioned a third type that was intermediate. The red cells are called *slow-oxidative*, the intermediate ones are *fast-oxidative/glycolytic* and the white cells are *fast-glycolytic*. The first sort (red) are rich in mitochondria and therefore have a high oxidative (aerobic) capacity, but a low glycolytic (anaerobic) one. Slow to contract and generating less stress, they have high endurance and can continue contracting for long periods of time. Fast-oxidative/glycolytic cells (intermediate) are faster, generate more stress, have moderate oxidative and glycolytic capacities, and high endurance. Fast-glycolytic cells (white) are the fastest, generate the most stress, have low oxidative and high glycolytic capacities, and a low endurance. All of the muscle cells within a motor unit are of the same type; thus slow, intermediate, and fast motor units have high, middling, and low endurances, respectively. Contraction in the three muscle types progresses in force from slow to fast cells, and when an animal changes gait from a walk to a trot to a gallop, the more powerful motor units are brought into action.

Different muscles often predominate in one of the three cell types. The composition varies among species according to their specializations. For example, the breast meat of a chicken is white and predominates in fast oxidative cells; duck breast muscle is red, evidence of their many slow oxidative cells. Chickens belong to a group of birds called galliforms that includes game birds like the partridge, noted for rapid takeoffs. This ability requires rapid and powerful muscle contractions but not endurance—they are sprinters, not endurance fliers. Ducks, in contrast, do not specialize in fast takeoffs, but are known for long migratory flights, hence endurance.

The physiological differences between sprint and endurance strategies are best illustrated on the sports field. World-class male sprinters can run one hundred meters in under ten seconds. They complete the entire event anaerobically, so endurance training is of no consequence. Sprinters' primary interests are in achieving maximum acceleration. For this they need maximum power, meaning large leg muscles. To build up their muscles, sprinters do a lot of weight training, lots of short fast runs, and explosive starts to improve acceleration. These exercises are largely anaerobic.

Sprinters do not just have muscular legs but also have well-developed upper bodies. The reason for this puzzled me until I had a chat with Andy Higgins, head track and field coach of the University of Toronto, who coaches athletes for the Olympics. He explained that when a sprinter explodes out of the blocks and accelerates down the track, he pushes with all his strength against the ground, first with one leg, then with the other. In doing so, his body twists one way and then the other; so if he did not have a sturdy upper body to resist these twisting forces, he would swing so violently from side to side that it would impair his running. Many female sprinters lack the strong upper bodies of their male counterparts and suffer from this problem.

Marathon runners, in contrast to sprinters, perform their event aerobically and need endurance, not acceleration. Their training program comprises aerobic exercises—mostly long-distance running—where the muscles perform at a rate where they can be supplied with sufficient oxygen for their needs. Endurance training increases heart size, hence blood output, increases the volume of blood, builds up the number of capillaries supplying the muscles, and adds more mitochondria in all of the muscle cells. There is also evidence that some fast-glycolytic muscle cells become transformed into the more endurable fast-oxidative/glycolytic type.

We have seen that large animals run faster than smaller ones, but how do their costs of transport compare? When we think about the cost of transport the price of airline tickets comes readily to mind because flying is what our society does a lot of. The introduction of large aircraft like the Boeing 747 in the early seventies reduced airfares primarily because it is much cheaper to fly four hundred people from New York to London in one large aircraft than in two or three smaller ones. Animals spend much of their time moving about in search of food, and transport costs probably constitute a sizeable portion of their total energy budget.

The subject of transport costs has occupied the attention of numerous biologists over the last few decades. They have constructed a general picture of the relation between metabolic costs of terrestrial locomotion and body size. Many experimenters have put animals on treadmills and assessed their transport costs by measuring how much oxygen they consumed. One of the more surprising findings is that the energy cost of transport, expressed in joules per kilogram per stride at the same equivalent speed, is pretty well constant over a wide range of animals, independent of body mass. The energy cost at the trot/gallop transition is 5.3 joules/kilogram/stride. This means that a horse and a small dog have to expend 5.3 joules per stride when they are running at the trot/gallop transition to transport each kilogram of their body mass. Although their cost per stride is the same, the dog takes more strides per second at an equivalent speed than does the horse. It follows that the dog's transport costs are higher. What is more, smaller animals probably have to use faster muscle cells at a given speed than do larger animals because they must stride more rapidly. Since faster contracting muscle cells cost more to operate than slower ones, this exacerbates their situation. Small animals therefore have higher transport costs than large animals, which coincides with their higher basal metabolic costs. Little wonder that their lives are a frenzy of activity that eventually drives them to an early grave—which is the subject of the next chapter.

FROM CRADLE TO GRAVE

When an old and decrepit Jack Crabb, in the movie *Little Big Man*, croaked to his interviewer that he had witnessed Custer's last stand, most of the audience probably did a quick calculation to see if his claim was feasible. The scene was set in the 1960s, and the Battle of Little Bighorn was in 1876, so it is conceivable that a young man could have witnessed the demise of Lieutenant Colonel George Armstrong Custer and lived to see the birth of the Beatles. If Jack Crabb had been sixteen at the time, he would have been 108 in 1968, which is a ripe old age but an attainable one. It is an intriguing thought that there just *could* be a living witness to the days of the great Sioux nation, but a little later on in the movie the irascible centenarian claims to be 121 years old, pushing credibility beyond the brink.

This movie scene teaches two important object lessons in the study of longevity: Some individuals of some species *do* live to a great age, and certain claims to antiquity are greatly exaggerated. The fine line between fact and fantasy is often difficult to draw, and although claims of advanced years are sometimes verifiable, many more are not. The major problem here is that records of individual ages, especially those dating back a century or more, are not always documented. The situation is bad enough when investigating human ages, but is worse when it comes to other animals. We have probably all heard accounts of how ancient a certain person or animal was, but most of these have no more veracity than the proverbial fishermen's tales. Old Mrs. Hubble, who lived next-door-but-one to the house where I was born, was said to be over one hundred years old. She did not look *that* old to me, but verifying her exact age was as elusive as establishing whether she really did take a cold bath every day—purportedly the secret of her longevity.

Aside from the Mrs. Hubbles of the world, it is a verified fact that our species can live to be about 110, and history abounds with records of such venerable old ages. Is this contrary to the fact that people died much earlier in the old days? Only a century ago most people were lucky to reach their fifties. As with any other variable, like height and weight, individuals cluster around the average mark, but there are always a few who are exceptionally tall or outstandingly heavy. Those of our ancestors who survived childhood and made it into adulthood may have been lucky enough to avoid dying on the battlefield or contracting cholera, but most would have died before reaching a venerable age. A very few, though, would have avoided succumbing to the vicissitudes of life and reached a century or more. But they tell us nothing whatsoever about the life expectancy of the rest.

Life expectancy is the average number of years to which a person of any age can expect to live. If you apply for a life insurance policy the agent will set the premium at a rate that reflects how long you are likely to live—and pay the premium. The agent's crystal ball is an actuary table, which predicts life expectancy based on your present age, occupation, general health, and life-style. A twenty-year-old letter-carrier who enjoys good health and who does not smoke obviously has a longer life expectancy than a fifty-year-old stockbroker who smokes heavily and suffers from work-related stress. Actuary tables are based on data accumulated over many years, and although you might be one of the lucky few who is going to live into your late nineties, chances are that you will die at the age of approximately seventy-three (if you happen to be male), as suggested by the table.

People have been concerned with predicting the age of death for well over a century. Now that we are both in a black frame of mind, let me beguile you with a grim statistic discovered in 1825 by the English actuary Benjamin Gompertz. He found that, after the age of about thirty, the likelihood of dying doubled every eight years—a statistic that has not changed. This phenomenon, where the chances of dying increase as an individual passes her prime, is described as *senescence* and applies to most living organisms. Senescence means that as a body ages its ability to cope with injuries and disease diminishes. A bad fall or a case of influenza may be of little consequence to a twenty-year-old, but can prove fatal to an octogenarian.

The life expectancy at birth for females in the United States was forty-seven in 1900, but this had almost doubled to about seventy-five by 1988. Similar improvements occurred throughout the rest of the western world and were primarily due to marked reductions in infant deaths and deaths during childbirth. To illustrate this graphically we can plot the rate of deaths (in this instance the number of individuals dying per 100,000) against age. The graph for 1900 has a sharp peak corresponding to infant deaths and falls sharply by early adolescence. It then climbs to a plateau during child-bearing years, rises to another peak at about seventy years, then tails off toward one hundred years. The graph for 1985, in contrast, has a smaller peak that corresponds to infant deaths, and this is followed by a period of almost zero deaths to the age of about twelve. The graph then climbs to a narrow peak at eighty-five years before tailing off, as previously, toward one hundred years. Survival curves are also available for many animals, especially the ones whose age can be determined. Fishes, for example, can be aged by counting the annual growth rings on their scales. The same is true for other structures like the shells of mollusks and turtles, and the waxy ear plugs of whales. When data are adequate, as with some commercially important species, meaningful survival graphs can be plotted for wild populations. For a vast majority of spe-

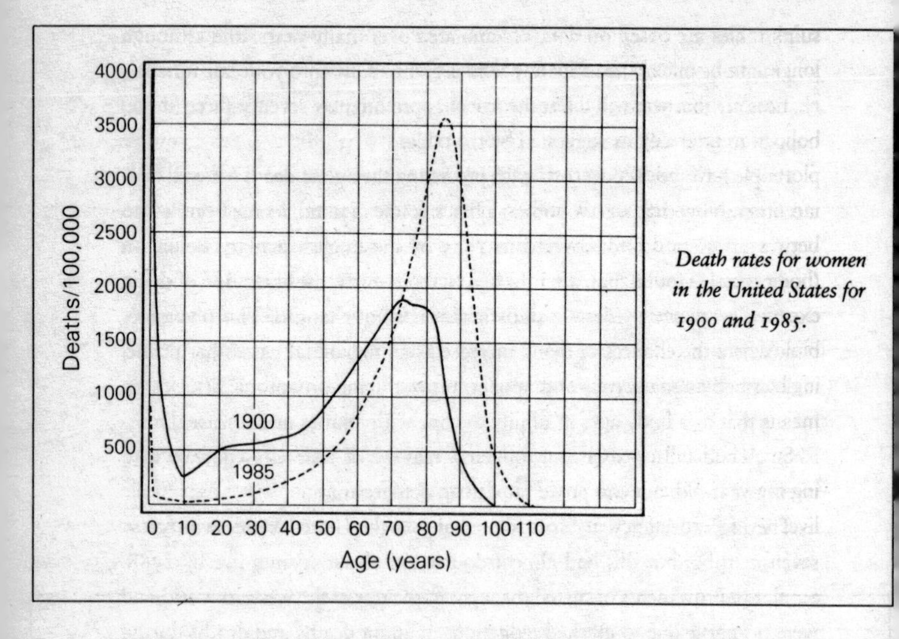

Death rates for women in the United States for 1900 and 1985.

cies, though, we rely on zoo records and personal recollections; so the best we can hope for is an idea of the *life span* of a species, that is, the age at death of the oldest survivor.

Most of us have an idea of the life span of common animals through our experience with family pets. Gerbils live a couple of years, mice about three years, rats five or six, cats and dogs upward of twenty years and sometimes more. Those familiar with horses know that a twenty-five-year-old would be considered ancient, but horses can live into their forties. Among zoo animals, gorillas live to their late forties, hippos and rhinos to about fifty, and elephants have been reported to live between sixty and seventy years. An obvious size-ordered trend here shows that larger animals live longer than smaller ones. This not only applies to mammals but also to birds, though the patterns are not so clear. Canaries and budgerigars live for ten to fifteen years, pigeons for thirty, the red-and-blue macaw for thirty-eight, and the greater sulphur-crested cockatoo for fifty-six years. Notice that birds live much longer than similarly sized mammals, a point to which we will return later.

The relation between longevity and body mass, like that between metabolic rate and body mass, is a power function. When the two variables are plotted logarithmically, two straight-line graphs are obtained. The mammalian line lies beneath the avian one. The gradients for the two graphs range between 0.19 and 0.29, so longevity is scaled approximately according to (body mass)$^{1/4}$, abbreviated $M^{1/4}$. The fact that metabolic rates rise and life expectancy shortens when body mass falls did not escape the attention of biologists, nor did the fact that certain other physiological variables, including heart rate and breathing rate, scale at approximately the same rate ([body mass]$^{1/4}$).

Small animals have high metabolic rates, high heart rates, and high breathing rates, and they reach sexual maturity and breed at an earlier age. Their lives are a flurry of activity compressed into a short time. As far as a mouse

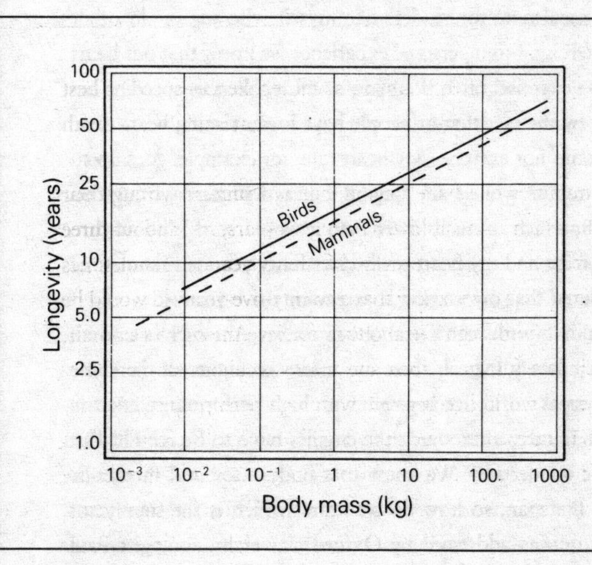

The relationship between longevity and body mass for birds and mammals (data are logarithmic).

is concerned, however, its three-year life is equivalent to an elephant's sixty-year span, as reflected in the tempo of its physiological functions. For example, the mouse's heart beats about seven hundred times a minute for two and a half years, which amounts to about 900 million beats. An elephant's heart beats about thirty times a minute for sixty years, amounting to about 950 million beats. Calculations like these have led to the idea of physiological time, whereby an animal's time scale is based on the rates of physiological processes occurring in its body, rather than on calendar time. A three-week period in the life of a mouse, during which time a female can conceive and have a litter, is therefore equivalent to twenty-two months in an elephant's life, that being its gestation period.

Since endothermic animals have roughly the same number of heartbeats during their life, the somewhat fanciful notion has arisen that there is approximately the same amount of "life" in each animal's heart—like a battery—and once they have had their allotted beats they die. I personally do not think there is much merit to the idea, but I am not a physiologist, so I will add a parenthetical note for readers like myself who jog or do other forms of physical exercise. From personal experience we know that our heart rates quicken during exercise, often doubling or more. Yet this speeding up is compensated for by the fact that fit people have lower resting heart rates than do people who do not exercise. My heart rate, for example, goes up to about 150 beats a minute while I am jogging, but as I sit here writing this chapter it is only 58, which is much lower than it would be if I did not exercise. So if I am wrong and my heart really does have a limited number of beats, I can rest assured that my jogging is extending the warranty.

The idea that animals with high metabolic rates drive themselves into an early grave appears entirely logical; there are many analogues of the same principle in our physical world. Racing cars with high-performance engines are good for only one race, after which the engines have to be rebuilt. But what about the size connection? We know that body mass *and* metabolic rate correlate with life span, so how do we know which is the significant factor? This question was addressed by Oxford University zoologist Paul Harvey and his colleague Mark Pagel who looked at the life-history data of several mammals. By factoring out metabolic rate and then body mass, they showed that body mass, not metabolic rate, was the significant variable. Long life spans therefore correlate with large body mass, not with low metabolic rate. Their findings also revealed several other life-history variables besides life span that correlate with body mass: maximum reproductive life span (the time during which an animal can reproduce), number of offspring per litter, and number of litters per year.

The relation between life span and body mass is well established, at least for mammals and birds. When the two variables are plotted using logarithmic data, a straight-line graph results. As with any generalization, there are always exceptions, as evidenced by the points that lie off the graph. One most notable exception is our own species, whose life span is considerably longer than it should be for animals of our body mass; horses, for example, weigh four or five times more than we do but live only about half as long. We have the longest life span of any endotherm. Why? Our medicines and the good care we take of ourselves can be ruled out because this does not change the maximum age to which some individuals of the species live.

One of our most outstanding features that sets us apart from all other animals is our inordinately large brain. Some say that this is the reason we live so long. The essentials of the argument are that a large brain is better equipped to maintain the constancy of the internal environment—body temperature, blood pressure, chemical balance, and the like—and this constancy extends our life span. A crude analogy: if you regularly service your car, keeping all the fluid levels topped up and the tire pressures correct, the car will last longer.

Why a small brain should not function just as well in performing the relatively menial tasks of maintaining the body's internal environment is a question that has occurred to a lot of skeptics of the big-brain argument, myself included. The argument's defenders point out that the correlation between brain mass and life span is closer than the one between body mass and life span. Therefore, when life span is plotted against brain mass rather than body mass, the graph is much tighter and has fewer outlying points. To counter this, the skeptics point out that this graph merely shows that brain mass is a better indicator of body size than is body mass. Consider, for instance, the wide range in body mass among adults of our own species. Though people can weigh from under 100 pounds (45 kilograms) to over 300 pounds (140 kilograms), their brain mass varies from about 2 to 4 pounds (1 to 2 kilograms).

The danger of making cause-and-effect arguments like the one between brain mass and life span is that the correlations between the two could be entirely spurious. Spleen mass, for example, is more closely correlated with life span than brain mass, but few biologists would want to suggest a causal link between the two. If the correlation between brain mass and life span is spurious, so too might be the correlation between body mass and life span, which leaves us with no correlations at all. Where do we go from here? What *is* the likely reason for long life spans? The best way to find an answer to this question is to look at animals that live longest and see if they have anything in common.

Among vertebrates, tortoises enjoy the longest life spans, though some of the longevity claims are probably more fanciful than factual. The most colorful is the royal tortoise of Tonga, which is said to have belonged to Captain James Cook. His first voyage to the Pacific began in 1768; Cook died in 1779 (at the hands of Hawaiian warriors), and the tortoise died in 1966. Assuming that his tortoise—if indeed he ever had one—was less than a year old when he got it, the animal would have been between 187 and 198 years old when it died, certainly the oldest known tortoise. Another ancient tortoise purportedly wandered the grounds of a British fort on the island of Mauritius until its untimely death in a fall at the age of over 150 years. Regardless of the veracity of these two reports, we do have reliable data for tortoises that lived for about 70 years. Incidentally, the fact that tortoises, being reptiles, have relatively small brains, is a count against the big brain–longevity argument.

Crocodiles and alligators also live long lives, as evidenced by the large size attained in the wild. In captivity they have lived more than 50 years. Some fishes live to ripe old ages too, including rock fish (over 100 years), sturgeon (82), catfish (60-plus), halibut (60-plus), and eels (55). Elephants live to a great age; the smaller of the two living species, the Asian elephant, reaches ages of between 60 and 70 years, whereas the African elephant may not live quite as long. Information for the largest of living animals, the cetaceans, is scant, mainly because it is difficult to assess their age, but data do exist. In one study on the sei whale, it was established from whaling boat data that they lived to at least 70 years while another study reported ages beyond 80 years. Some birds have similarly long life spans. The longest-lived birds include condors (70-plus years), parrots (70), the eagle owl (68), and the white pelican (51). The largest of all living birds, the ostrich and its flightless relatives, live only about 28 years. The herring gull, which is only a fraction of the size of an ostrich, has a surprisingly long life span of about 41 years.

While reports of tortoises living for nearly two centuries are probably unfounded, we have reliable evidence that some invertebrate animals have lived even longer. The oldest of them all is a rather unpretentious-looking clam with an unromantic name: ocean quahog. Like many other mollusks, its shell is marked by concentric growth rings, and from observing living specimens we know that these rings are annual. Ring counts reveal that this humble mollusk lives for over 200 years. There are other long-lived clams, notably the freshwater mussel and the Pacific giant clam, which live for more than a century. Aside from their remarkable longevity, these animals show no signs of senescence, with no increase in mortality during their advancing years. They continue growing, albeit slowly, and keep reproducing throughout their long lives—which may explain the origin of the expression "as happy as a clam." Other examples of long-lived invertebrates include sponges (50 to 100 years), anemones (80 to 90), and lobsters (50; estimated, not recorded).

What do long-lived species have in common? What could a clam and a condor possibly have in common with a tortoise and a sturgeon? The answer might lie in the pages of Lewis Carroll, but we shall seek an answer in the works of one of his contemporaries, Charles Darwin.

Darwin's theory of evolution by means of natural selection is elegant in its simplicity. Offspring are similar to, but not identical with, their parents. Consequently no two individuals are exactly alike, except identical twins. Each species produces far more offspring than can possibly survive. As the offspring are not exactly alike, it follows that some individuals will have features that give them an advantage over others. These advantages give the individual a better chance of surviving and therefore of leaving more offspring. Such advantaged individuals are referred to as being *fit*. The term *fitness*, as used here in the evolutionary context of producing more offspring, has nothing to do with the usual meaning of the word. Since the offspring of the advantaged individuals inherit some of the favorable features of their parents, they too have an improved chance of survival. Darwin termed this process—the selection of advantaged individuals—*natural selection*. Favorable features (i.e., favorable genes) are said to be selected for, and are passed on to the next generation. Favorable genes eventually spread throughout the gene pool of the population. The gene pool is the sum total of all the genes of that population. It is through the action of natural selection that populations of a species can adapt to changes in the environment.

A good example of natural selection in action was provided by the effects of the severe El Niño of 1982–1983 on the birds of the Galápagos Islands. El Niño is the name given to the periodic warming that takes place in the eastern Pacific Ocean, often with severe environmental effects. During the 1982–1983 El Niño, the most severe of the century up till then, the Galápagos archipelago experienced an unusually wet rainy season, followed by several years of drought. Most of the usual food resources of the indigenous birds were destroyed.

Among the many casualties were the finches that bear Darwin's name, including the cactus finch. The sparrow-sized cactus finch normally feeds on the flowers of the *Opuntia* cactus (prickly pear), a plant that grows to the size of trees. Aside from feeding on these flowers, available only during the wet season, cactus finches also feed on caterpillars, spiders, and small seeds. The birds breed during the rainy season, when food supplies are plentiful, but food becomes scarce during the dry season. This is the time of highest mortality, when the birds are forced to exploit other food sources to survive. Some individuals break open the spiny fruits of the cactus and feed on the outer casing that surrounds the seeds. Others crack open the hard seeds and eat the kernels. These same individuals also strip the bark from trees to catch insects. The shape of the beak varies among individuals of the species; some cactus finches have short deep beaks while others have longer and more slender ones. Individuals with long slender beaks generally feed on seed cases, but only the ones with short deep beaks can crack open seeds or strip off tough bark.

During the heavy rains of the El Niño many *Opuntia* trees became waterlogged and blew over in the high winds. The only food available to the cactus finches during the ensuing drought were the insects living beneath the bark and inside the sun-baked pads of the fallen cactus trees. Seeds belonging to other species of plants were available, but these were much tougher to crack than the *Opuntia* seeds, and could be tackled only by birds with deep beaks. Individuals with slender beaks were at a serious disadvantage and most of them starved to death. Casualties were high for all birds, but the majority of

*Variation in beak size and shape among adult cactus finches (*Geospiza conirostris*).*

survivors were the ones with deep beaks. As a consequence, for the next several generations the cactus finch populations predominated in individuals with deep beaks. The change was not permanent, though. When things started to return to normal about 1986, slender-beaked individuals were no longer at a disadvantage and began appearing again in the population. Natural selection holds the key for understanding why some species become senescent within a few years while others enjoy long lives.

Senescence can be thought of as a decline in the general health of an individual past its prime. With apologies to readers like myself who are on the wrong side of 50, I will relate these changes to our own species. In our teens and early twenties we were invincible, participating in the most strenuous of sports with little or no after-effects. If we did get hurt, our injuries soon repaired. And it would have taken a serious malady to make us cancel Friday-night plans. By our late twenties and early thirties we become more aware of stiffness following a hard game. Stubbed toes and minor sprains begin to niggle, and a winter virus might send us to bed for a day. Forty-year-olds definitely have to look after themselves. If they play or work too hard, they can expect pulled backs, aching joints, and inflamed ligaments. This is about the time of life when you become aware of the mileage on your equipment. If you have not encountered poor eyesight, arthritis, hernias, gallstones, hemorrhoids, dyspepsia, and back pain, do not despair—you have them to look forward to in your fifties! Chances of contracting serious diseases also increase (remember that the probability of dying doubles every eight years). None of this sounds promising, but we might be lucky and slip graciously into our dotage.

Why do all those bad things start happening when they do? And what is to stop them from happening much sooner? From an evolutionary perspective, an earlier commencement of senescence would reduce an individual's evolutionary fitness because it would reduce its chances of producing offspring. The other side of the coin is that postponing senescence until *after* the individual's reproductive life has passed increases its fitness. Any feature that postpones senescence during an individual's reproductive life will be an advantage and consequently will be selected for, assuming that there are no accompanying negative affects. What sort of features can delay senescence? Consider all the repair and replacement mechanisms taking place in the body on a continuing basis. With the exception of nerve cells and muscle cells, which do not appear capable of division, our other body cells are being replaced all the time. Red blood cells, for example, last only three or four months. Senescence-delaying features therefore involve those genes that enhance cell replacement. Cell divisions sometimes result in copying errors in the DNA; such errors are called gene mutations. Most mutations are harmful, and if allowed to accumulate they lead to serious malfunctions, including cancers. Indeed, a whole class of genes, called tumor-suppressing genes, repairs defective DNA. These genes are clearly another group of senescence-delaying features.

The advantage in having these features operational during an individual's reproductive life is that they heighten the chances of producing offspring. Once the individual is past its reproductive stage, these features are of no significance to its fitness. Fortunately the repair and replacement mechanisms that worked during the reproductive period will operate afterward, but maybe not as well. New problems can arise and there might not be a repair or replacement mechanism available to take care of them simply because such features have not been selected for. This can be bad news for the individual and may be why certain diseases, like many cancers, occur predominantly in middle and later years.

The correlation between longevity and body size is well established and is also correlated with breeding rates. Animals with short reproductive lives tend to breed young, produce numerous offspring, and have short life spans. Those with long reproductive lives breed later in life, produce fewer offspring, and have long life spans. Some shrews, for example, give birth when they are only seven weeks old, have large litters and short lives. The two longest-lived mammals, elephants and humans, do not become sexually mature until about eleven years old, they produce few offspring, and have the longest documented reproductive lives of any mammal—up to thirty or forty years. Tortoises remain reproductively active throughout their lives; the painted turtle even lays more eggs at the age of twenty-five than at the age of ten. There is also evidence that long-lived birds have long reproductive lives too. This continuum between small, short-lived animals like mice that produce large numbers of offspring, and large, long-lived ones like elephants that leave far fewer offspring can be viewed as a trade-off between investing energy in reproduction or in growth. Indeed, longevity may be more closely linked to breeding strategies than to body size. We tend to make a causal link between longevity and body size because they are so strongly correlated, but this is not necessarily the case. Snow, for example, is correlated with the low temperatures of winter, which are also correlated with shortened days, but it would be wrong to assume a causal link between snow formation and day length. While some aspects of large size, such as freedom from predators, are obviously causally related to longevity, we have to be careful not to assume that large size, per se, is the underlying cause of longevity. The causal uncoupling of body size and longevity helps explain some discrepancies such as why cats usually outlive large dogs and why humans outlive horses.

The correlation between life span and breeding habits has experimental evidence to support it. Take for instance the breeding experiments with geneticists' favorite animal, *Drosophila*, the fruit fly, which can be seen flying around fruit bowls during the summer months. Taking about ten days to grow from an egg to a sexually mature adult, they have an average life expectancy of about one month; their maximum life span is about two months. In one experiment, one-day-old flies were divided into two groups and put into separate bottles provided with food. One of the groups was allowed to breed freely, without any interference from the investigators. In the other group, eggs were collected at set times and transferred to a fresh bottle for raising into the next generation. During the experiment's first phase freshly laid eggs were collected on the twenty-eighth day. These were eggs of twenty-eight-day-old adults—which are fairly senior individuals—and many of the adults' companions had already died of old age. When the eggs hatched and matured to adults the procedure was repeated: eggs were collected again from twenty-eight-day-old adults. This collection cycle continued for a couple more generations to build up a stock of older-breeding flies. Then, for the next few generations, eggs were collected on the thirty-fifth day, then on the fifty-sixth day, and finally on the seventieth day. Now a seventy-day-old *Drosophila* is positively ancient—the equivalent of breeding from one-hundred-year-old humans! When the experimenters compared females from the late-breeding population with females that had not been selectively bred, they discovered that the former group's life span had lengthened by as much as twenty days. Hence longevity can be increased by selectively breeding from older individuals—strong support for the link between longevity and long reproductive life.

I mentioned that a seventy-day-old *Drosophila* laying eggs is the equivalent of an old woman bearing a child, but this is not strictly true because of the differences in the age of their eggs. Female mammals are born with their full complement of eggs* whereas other animals produce new eggs during their reproductive life. Therefore, when a forty-year-old woman becomes pregnant, her baby develops from a forty-year-old egg, which probably explains why the incidence of abnormal babies increases with the mother's age. Males, in contrast, produce new sperms throughout their reproductive lives, which explains why elderly men are able to sire healthy children. In recent medical developments, childbirth in post-menopausal women involves the implantation of young fertilized eggs from another woman.

Another side to the question of longevity is the role of risk factors. Simply stated, it is pointless for an animal to evolve internal features enabling it to have a long reproductive life and a long life span if it is likely to be eaten! Are animals with long life spans especially well endowed to cope with these external factors? Again the answer appears to be yes. Elephants, because of their large size, have no natural enemies besides humans, and the tortoise's thick shell gives it a large measure of protection against predators and other hazards, as I learned at an early age. I grew up in London during the latter days of World War II, and when the war was over we youngsters used to play in bombed-out buildings. On one of our jaunts my older brother found a tortoise wandering around a burned-out house. Its shell was badly scarred from the fire, but it survived and lived as our pet for over thirty years. Like

*The bush baby (*Galago*) may be able to produce new eggs during its reproductive life (C. E. Finch, personal communication, January 1994).

elephants, condors and other long-lived birds of prey appear to have no enemies. The fact that birds can fly away from danger explains why they have much longer life spans than mammals of similar size. Derek Pomeroy, a zoologist at Makerere University in Uganda, studied mortality rates in birds and found that those that spent most of their time in the air, like swifts, had the lowest mortality rates. Bats, predictably, have longer life spans than other mammals of similar size, while ostriches and their flightless relatives, and penguins, live rather short lives compared with flying birds.

The risk-factor idea is a promising one, and opossums have provided supporting evidence. The opossum, a cat-sized mammal belonging to the same group as the kangaroo, is the only marsupial that lives in North America. Marsupials have lower metabolic rates than placental mammals and some tend to be somewhat slower moving (there may be some correlation between the two). This is certainly true of the opossum. Just look at their alarmingly high rate of highway accidents; some roads in the southern United States are splattered with their earthly remains during the spring. They are equally unfortunate in their encounters with predators, even though they are nocturnal, which supposedly reduces the risk of being attacked. According to evolutionary theory, opossums should breed young (before it's too late!), produce many offspring, become senescent, and die soon. And this is exactly what they do. They can breed as young as six to eight months old, have two litters a year (each litter averages eight and sometimes reaches thirteen) and live a mere five years. The raccoon, in contrast, an ecological equivalent of somewhat similar size, breeds when one or two years old, usually has a single litter each year of between two and five offspring, and has a life span in the wild of about sixteen years.

An island 5 miles (8 kilometers) off the coast of Georgia in the southern United States has been separated from the mainland for about four thousand years, and, fortunately for the resident opossums, it is free from their mammalian predators. This opossum haven is so safe that the opossums sometimes wander about during the daytime and do not even bother to retreat into burrows to sleep. It should come as no surprise to learn that these island opossums have fewer offspring than their mainland counterparts, do not become senescent so rapidly, and have longer life spans—totally consistent with the evolutionary theory of senescence.

Before ending our morbid sojourn to the end of time, we should look at the smallest of living things, the protozoa and bacteria. These microscopic organisms are found in every conceivable niche on earth and are responsible for many of our debilitating diseases. They usually reproduce asexually simply by splitting in two. The cells of our multicellular bodies do the same thing too but with a fundamental difference; our cells appear to become senescent whereas protozoan and bacterial cells never grow old.* If endothelial cells (those that line the inside of blood vessels) are taken from our body, placed in a nutrient medium in a petri dish, and incubated, they undergo repeated cell division. Eventually a mat of cells, one cell thick, will form on the bottom of the petri dish. Yet as soon as they touch the sides of the dish, they stop dividing. If cells are removed and placed in a fresh petri dish, however, they start dividing again and will continue to do so until they touch the sides of the dish. This process can be repeated about fifty times, but then the cells die because they have become senescent. Many criticisms, however, have been made of these experiments. First, the cells become transformed during the process, and those at the end of the experiment are not exactly like the original cells. The possibility exists that cell death may be attributable to problems with the nutrient medium rather than with the cells themselves. It has also been argued that far more cell divisions occur during the experiment than would occur in the body, even in the most actively dividing cells, so the results have no relevance to senescence anyway. Be that as it may, it is interesting to note that the number of repetitions of the division process is directly related to the life span of the donor species. Cells removed from mice therefore undergo about ten cycles, compared with fifty or sixty for humans. If the same experiment were repeated with bacteria or with protozoans, they

*If single-celled organisms *did* grow old they would become extinct because their reproduction involves the whole body splitting in two; multicellular organisms, in contrast, have separate reproductive cells.

would go on dividing forever and never become senescent. The material making up the bodies of these organisms is continuously being replaced and renewed; the organisms themselves never grow old. One of life's many ironies is that these lowly organisms, some of which can strike us down, enjoy immortality.

GIANTS—MODERN AND ANCIENT

EQUESTRIANS MIGHT TAKE the large size of their mounts for granted, but for those who are rarely seen in the company of horses, their size can be quite daunting. Viewed from up close, a horse is a substantial animal. Patting a horse is like slapping your hand against a brick wall. If the horse gives you a friendly nudge in return, it is almost enough to knock you off your feet. The average horse weighs about 1000 pounds (450 kilograms), all of which is supported by the skeleton. Tetrapods rely on a structure analogous to that of a bridge: The vertebral column forms the arch and the fore and hind legs transmit the forces of weight to the ground. Tetrapods, from mice to sauropods, have basically similar skeletons, so we can use the horse as an example. The vertebral column, the central supporting structure of the skeleton, comprises a large number of vertebrae joined together to form a stiff but flexible rod. The individual vertebrae vary in shape according to their position along the column; hence we can distinguish between, say, a caudal vertebra from the tail and a cervical vertebra from the neck.

Each vertebra consists of a short cylinder of bone, the centrum, surmounted by the neural arch, through which the spinal cord passes from the brain. The neural arch is usually drawn out dorsally (upward) into a prominent bony process, the neural spine, to which muscles and ligaments are at-

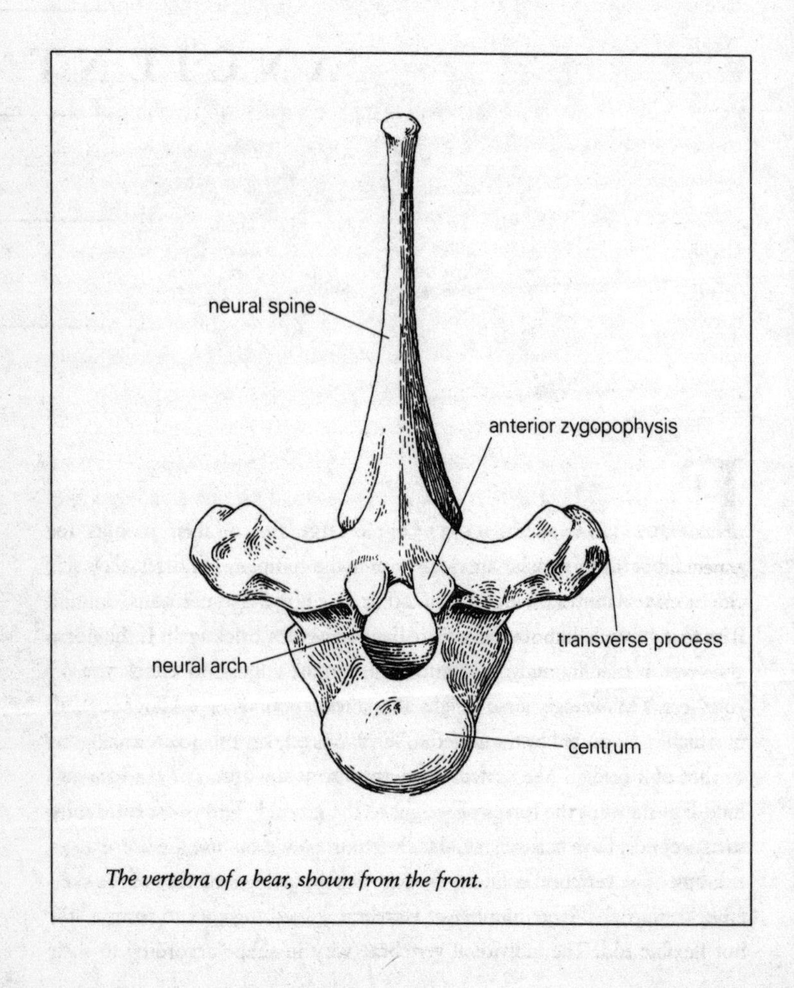

The vertebra of a bear, shown from the front.

tached. Ligaments, made of a tough protein called collagen, are the sinews that connect bones together. At about the level where the neural arch joins the centrum are a pair of processes, the transverse processes, to which the ribs are attached or articulated. Adjacent vertebrae articulate with one another by paired processes, fore and aft, called *zygapophyses*, which lie at the level of the neural arch. The articular surfaces of the anterior (front) zygapophyses face upward and inward and correspond with those of the posterior (back) zygapophyses, which face downward and outward. The anterior and posterior surfaces of the centra also articulate, and in many animals, including many dinosaurs, the anterior surface is convex while the posterior surface is concave, forming a ball-and-socket joint. The individual vertebrae are held together by intervertebral discs, which are tough but compliant, so they provide a small amount of give between adjacent vertebrae. The vertebrae are also held together by ligaments that bridge across two or more vertebrae at a time. Muscles also help. For example, numerous small muscles crisscross between adjacent neural spines and transverse processes. Like ligaments and intervertebral discs, muscles are compliant, and, since they are contractile, they can alter the animal's posture.

The vertebral column is usually not straight as in the horse but arched, like a bow, between the pectoral (shoulder) and pelvic (hip) girdles. This curvature is maintained partly by the shape of individual vertebrae and how they articulate with one another, and partly by the tension in the ligaments and especially in the muscles. Muscles also bring about lateral as well as up and down movements of the column, and they are arranged in two major groups. The first group lies above the level of the transverse processes and straightens the arch. These numerous muscles are either short, like those spanning from one neural spine to the next, or considerably longer. When we have the misfortune to "put out" our back, usually by lifting something awkwardly, we have usually pulled one of these muscles. The second group of muscles lies below (ventral to) the level of the transverse processes, and complements the action of the first group and increases the curvature of the spine. Some of these muscles crisscross between adjacent vertebrae, while others span across the ribs, both externally and internally.

Abdominal muscles are also part of this group and are the most important ones for counteracting the tendency of the vertebral column to sag between the shoulders and the hips. Abdominal muscles are especially well developed in heavy, long-bodied animals. They are the main reason why a horse does not sag even when a heavy rider climbs into the saddle.

The vertebral column clearly provides the body with a stiff but flexible supporting beam, and its extensive musculature maintains stability and mo-

A simple beam (left)*, and an engineer's I-beam.*

bility. But a word or two about beams is needed before continuing with our account of the skeleton. The beam, a common engineering structure, is simply a horizontal device that supports a load. A plank of wood supported at either end on bricks is a beam, and if you stood in the middle of it, the plank would sag beneath your weight. It is not difficult to visualize the wood fibers on the plank's lower surface pulling apart as the plank bows down. These fibers are being loaded in tension. Conversely, fibers on the top surface are pushed together; that is, they are loaded in compression. Since the top of the beam is loaded in compression and the bottom is loaded in tension, it follows that both stresses decrease toward the beam's center. It also follows that a zone in the middle, the neutral axis of the beam, has no stresses at all. Since there are no stresses in the neutral axis, holes can be drilled through it without weakening the beam. Indeed, engineers use this strategy to reduce a beam's weight; hence the invention of the I-beam, where most material is concentrated away from the neutral axis.

Another common engineering structure is the column, a vertical device for carrying loads. If you lifted the plank of wood off the bricks, you could use it as a column to prop up a ceiling. When oriented vertically like this all the wood fibers are pushed together, and the column is loaded entirely in compression (this is not strictly true because the stress on the column causes it to bulge slightly, generating some tensile stresses).

The body's weight is supported by legs, which are connected to the vertebral column through their respective limb girdles: the pelvis for the hind legs and the pectoral or shoulder girdle for the fore legs. The pelvis has a firm bony union with the vertebral column, usually via bone-to-bone fusion, but the pectoral girdle is neither fused nor has direct contact with the vertebral column. Instead, the pectoral girdle is attached, primarily by muscles, to the rib cage, often with a good degree of mobility, as with humans' shoulders. In many animals, including horses, the pectoral girdle's movements help the forelegs during locomotion.

Legs have to do more than support the body; they must provide for movement, too, and sometimes compromises must be made between these two objectives. One such compromise involves the angle at which the limb segments are held with respect to the vertical. When the skeleton of a mammal, bird, or dinosaur is viewed from the front or back, the limb segments lie in a vertical line, described as the erect posture. The erect posture contrasts with the primitive sprawling posture of amphibians and most non-dinosaurian reptiles, where the fore and hind legs are held out at the sides of the body as if the animal were doing push-ups. But when skeletons with an erect posture

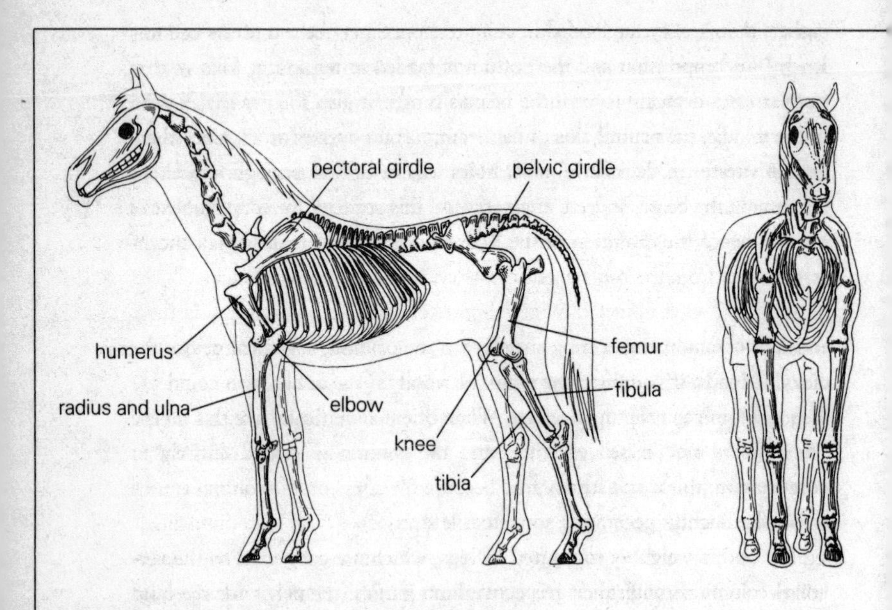

The skeleton of a horse. Notice that when viewed from the side the upper limb bones (humerus and femur) are angled relative to the others. (This angling of the limb bones is more marked in smaller animals.) Viewed from the front the limb bones are vertically in line.

are viewed from the side the limb bones are rarely oriented vertically. Instead, the humerus (upper arm) slopes backward and the radius and ulna (forearm) slope forward; the reverse holds for the hind limb where the femur (thigh) slopes forward and the tibia and fibula (shin) slope backward. The knee and elbow therefore point fore and aft, respectively. This puts added stress on the leg bones because they are being loaded in a way that bends them. (Most bones break during activities like skiing and roller-blading when bones are subjected to large bending stresses.) The obvious mechanical solution is to keep the bones vertical and to move with stiff legs. In this situation the bones are primarily loaded as vertical columns in simple compression.

But leg bones inclined to one another offer important locomotory advantages. The bones' inclination is maintained by tension in the muscles and ligaments—a system that provides for a natural springing action. This system enables tendons, ligaments, and bones to store energy, called *strain energy*, the same way that a spring does, which is important during locomotion. Having limb bones move relative to one another also speeds up leg movements because the speed of the foot is the sum of the speeds of the individual segments of the leg. What is more, the foot swings through a greater distance, as demonstrated by simple experiments with jointed popsicle sticks. Three popsicle sticks are immovably joined in a straight line by overlapping their ends and taping them. A second set is movably pinned together by thumbtacks. If both "legs" are swung through the same arc of the upper segment, the jointed one is able to extend the swing of its "foot" by moving its joints. The inclination of limb bones is therefore a compromise between weight support and locomotory performance.

The hollowness of bones compensates for the greater stresses on inclined leg bones. This strategy follows a well-known engineering principle that a hollow pipe can withstand greater bending stresses than a rod of equal weight. Thus steel pipes rather than steel rods are common in construction. Although horses are large animals, especially compared with our own modest stature, they could not be described as giants; the only terrestrial giants alive today are Asian and African elephants.

You will never see as many people at the elephant house of the zoo as you will find aping it up in front of "lesser" primates, but elephant watchers are an appreciative audience. There is something special about elephants, a primal attraction that is hard to pinpoint. Sheer size? Incongruity? An evocation of the remote past?

The last time I leaned on the railings to watch elephants I struck up a conversation with one of the former elephant keepers. He was now working with other animals in another part of the zoo, and during his coffee break he had come to visit his elephants. He told me how much he missed them. Elephant keepers always have elephant stories to tell; one recurring theme is that no matter how friendly and cooperative the great beasts may be, there is a good chance they will eventually try to kill their keepers. I have read of such fatalities in the past but assumed it was an accidental consequence of their large size. If a keeper happens to be against a wall when an elephant inadvertently swings around, he is going to get squashed. Not so, the elephant keeper assured me; they do it on purpose. He told me about an elephant who used to gather up pieces of bread with his trunk and make a trail for birds to follow. When an unsuspecting wild bird strayed too close, the elephant would bring down his foot and squash it flat! Accounts from Africa relate how elephants attack lions and trample them to death. Such tales of premeditated violence reveal little about the functional aspects of being a giant, except to explain why large animals are left alone by others, and why a primary factor in the evolution of gigantism is probably defense against predators.

The African elephant, weighing up to six tons or more, is the larger of the two species and the largest living land animal. Having to withstand such enormous forces of gravity must be tiresome, and this is apparent in the plodding way in which elephants move. Their actions are slow and deliberate, whether they are trudging about or picking up something with their trunk. When they walk they keep their legs straight, a stance they maintain when they run. Unlike other tetrapods, they are unable to gallop because this would put undue stresses on their bones. Although they sometimes rear up on their hind legs in the wild, to reach branches in trees, they do so with great deliberation. The long bones of their skeleton—the femur, tibia, and fibula in the hind limb and the humerus, radius, and ulna in the forelimb—remain vertical so their legs function primarily as columns. The bones are loaded as columns rather than as beams, and hence there is less advantage to their being tubular. Indeed, because it is desirable to maximize the cross-sectional area, to provide the largest area of bone material, their bones do not have large marrow cavities. If the bones remained hollow, larger diameters would be necessary to have the same cross-sectional areas.

The pectoral and pelvic girdles of the elephant, like the limbs, have a vertical orientation that reduces the bending stresses. The feet are held with the sole and palm raised off the ground, as if the elephant were standing on its toes. However, in life, the foot is cushioned by an extensive elastic pad so the weight is borne by the entire foot. Watch an elephant at the zoo and you will

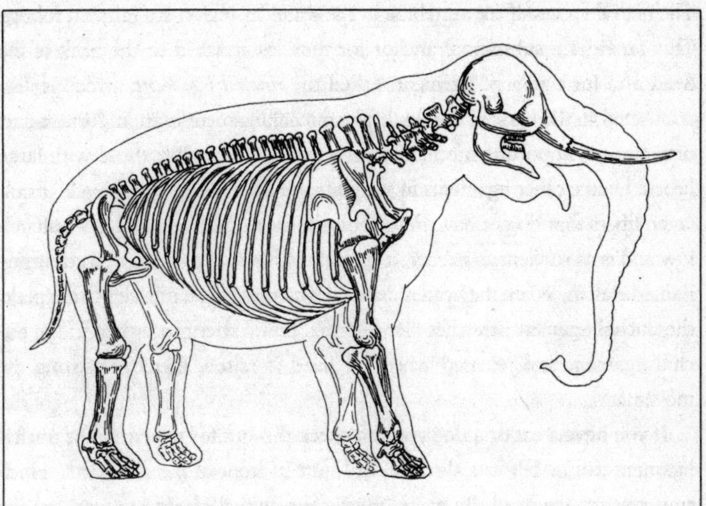

The skeleton of an Asian elephant, which is very similar to that of the African elephant. Notice that the orientation of the limb bones is essentially vertical, as are the limb girdles. These strategies reduce stress in the bones.

see that when its foot lifts clear of the ground the pad balloons out at the bottom. When it is put back on the ground, the pad flattens and spreads out like an under-inflated tire. This pad is remarkably resilient, like dense foam rubber, and absorbs much of the stress of the footfall, as with a well-cushioned jogging shoe.

The front feet are much larger than the hind ones, because they carry more weight than the back ones. The weight is attributable to the massive head—more than a ton for the African species—especially for males, whose tusks are larger than females'. Keeping the head up takes a lot of effort, and several anatomical features help in this task. The most striking feature is the neck's extreme shortness, which reduces the leverage that the head exerts on the skeleton. Like all other mammals, the elephant has seven cervical vertebrae. But these are quite short, as if they had been flattened from front to back.

The neural spines of the vertebrae in the shoulder region are tall and robust. This provides a substantial anchor for muscles attached to the back of the head and for the large ligament, called the *nuchal ligament*, which is also connected to the back of the skull. The nuchal ligament is an important part of the head-supporting mechanism of tetrapods, especially those with large heads. Unlike other ligaments in the body, which are white and made of collagen fibers that do not have much pliancy, the nuchal ligament is a buff yellow and is as resilient as rubber. It is made of fibers of a springy protein aptly named elastin. When the head is depressed by the ventral muscles of the neck, the nuchal ligament stretches like a spring. Strain energy is stored in the nuchal ligament and released when the head is raised, thereby assisting the movement.

If you have a cat or a dog you can check this out for yourself. The nuchal ligament can be felt as a discrete cord, just in front of the shoulders, which runs toward the head. By gently raising the animal's head you can feel the ligament slacken, and if you depress its head you will detect the tension increasing. Our cat's nuchal ligament is about one-quarter of an inch (5 millimeters) thick, but an elephant's is about as thick as your wrist. Since the neck is so short, a standing elephant is unable to reach the ground with its head. Even if it could do so, its tusks would prevent it from being able to pick up food. So the elephant has to rely on its trunk for gathering food. If this appendage is severely injured, as unfortunately happens in the wild when it gets caught in a poacher's snare, the elephant starves to death.

One implication of an elephant's low metabolic rate is that, in relation to its body weight, it requires less food than a smaller mammal. In absolute terms, a five-ton elephant consumes more food in a day than a one-ounce mouse, but five tons of mice would eat many times the quantity of food that a single elephant would. In spite of its low metabolic rate, an elephant has to eat such prodigious amounts of food that it spends about 75 percent of its time eating, both day and night. The main reason for this seeming gluttony is because its food is low in nutrients.

Plant food comprises two main parts: the contents of plant cells (cell sap) and cell walls. Cell sap comprises sugars, proteins, and starch and is readily digestible by enzymes secreted by vertebrates. (Enzymes are proteins that facilitate chemical reactions in living organisms, in this instance the digestion of large food molecules into smaller ones.) Tough cellulose walls, in contrast, cannot be digested by vertebrate enzymes; herbivores have to rely on bacteria in their gut to break down the cellulose. Except for rare food items like fruits, root tubers, and fresh young grass shoots, cell contents comprise only a small part of plant material. Herbivores must therefore consume a lot of indigestible fibrous material that has to be slowly broken down in the gut by bacteria. Gut capacity has the dimensions of volume, which means that gut size increases in step with body mass. The time the food remains in the gut also lengthens with body mass. Hence large herbivores can take more fibrous materials in their diets than smaller animals can—and they can retain such material longer to digest it more completely. With this ability they can exploit a wider range of plant foods. The environment abounds in low-quality plant foods (those with higher percentages of fiber); so giant animals, like the elephant, have a competitive edge over smaller ones. This is especially true when higher-quality food is less abundant, as in times of drought.

Although the African elephant spends up to eighteen hours a day eating, its life is not the frantic rush of a small mammal like a shrew. Elephants amble casually through life, and when it suits them they will take time out to wade into a water hole to cool off. With all the weight they have to haul around, lying down to sleep must be a sweet relief. But they must savor their joy in small doses because if they lie down for more than about an hour they risk serious damage to the side of their body touching the ground. Nerves can be injured, muscles bruised, and blood clots formed, the latter of which can kill them. The elephants' great weight makes them vulnerable to falls, and a tumble that would be no more than an inconvenience to a smaller animal could prove fatal. It is understandable that elephants avoid taking unnecessary risks and are careful of their footing when traveling over uncertain terrain.

Being large often means being tall, with all its associated physiological problems. The tallest of living animals, the giraffe, is 10 to 13 feet in height (3 to

4 meters—up to 5.5 meters in large individuals), with its head about 7 to 10 feet (2 to 3 meters) higher than its heart. For blood to reach its head and pass through all the fine blood vessels, the heart has to discharge blood at a pressure roughly one third of the atmosphere's pressure—about twice the pressure of the blood leaving human hearts.

Pressure is usually measured according to the height of a column of liquid that the pressure can support. One could, for example, measure the pressure available for watering a lawn by turning on a garden hose and hoisting the end of it so high that the municipal pressure could no longer force it out. Atmospheric pressure is equivalent to a 32-foot (10-meter) water column; but municipal water pressure is likely to be higher than this, so for the column to be shorter it is necessary to use a denser liquid than water. Mercury has been the liquid of choice for centuries. Atmospheric pressure is equivalent to a 30-inch (760-millimeter) column of mercury and, accordingly, blood pressure is measured in millimeters of mercury. The average pressure for blood leaving a healthy human heart during its contraction is 120 millimeters of mercury, abbreviated as 120 mm Hg.

Blood leaving a giraffe's heart is required to do more than reach the head; its pressure has to be high enough to pass through all the capillaries that supply the brain and other organs. To achieve this the blood pressure of a giraffe is about 200 mm Hg—probably the highest blood pressure of any living animal. The pressures are so high that they would probably rupture the blood vessels of any other animal, but two mechanisms prevent this. First, a giraffe's arterial walls are much thicker than those of other mammals, especially in the lower parts of the body where the pressures are highest. Second, the total tissue pressure, that is, the pressure exerted on the blood vessels by all the surrounding tissues, is kept high. This pressure is maintained by a tightly stretched thick skin that functions like the antigravity suits worn by pilots of fast aircraft.

I mentioned previously that heart rate declines as bodies become heavier. The giraffe, though, weighs a little over a ton and has a heart rate of about 66, a rate that is comparable to our own. Presumably this high rate in such a large animal is related to the need to generate high blood pressures.

That unpleasant dizzy sensation you may experience when you stand up quickly from a lying or sitting position (especially if you are not in good shape) is caused by a temporary fall in the pressure of the blood reaching the brain. Giraffes sometimes lie down to rest, and when they regain their feet they have to do so in stages, first squatting and waiting for a few moments before standing fully erect. This behavioral modification presumably allows the vascular system to stabilize. If a giraffe suddenly raised its head to its full height it would probably become faint. By the same token, when a giraffe is browsing or drinking its head is lower than its heart and the blood pressure increases. But the capillaries of the head are protected from this pressure increase by the constriction of the vessels supplying them with blood. It has been said that the reason why a giraffe splays its front legs while lowering its head is to reduce the difference in height between the head and the heart. But a giraffe *has* to do this to be able to reach the ground. Incidentally, long legs may have evolved in giraffes to minimize the distance between head and heart—if the legs were shorter the neck would have to be longer to give the giraffe an equally high reach.

Humans are not very tall, yet the height difference between the lower legs and the heart is enough to cause a problem in returning blood to the heart. This problem is primarily because blood in the veins—the vessels that return blood to the heart—flows at such low pressures. Why? The high-pressure arterial blood that leaves the heart has to be distributed to all parts of the body. The blood is forced through smaller and smaller vessels, from arteries to arterioles, and then through the extensive network of capillaries that supplies various tissues with blood. Capillaries, the smallest vessels, have diameters not much wider than the diameter of a red blood cell. The blood obviously leaves the capillaries at a low pressure and passes into venules, then finally reaches the veins. The blood from the legs now has to fight gravity and make its way back to the heart. During the blood's return the leg muscles' pumping action greatly assists the process. The way this works is that veins have valves which allow blood to flow only toward the heart; so when the leg muscles massage the veins, blood is squeezed toward the heart. The extreme discomfort that soldiers experience when forced to stand still on a parade ground is mainly due to the muscles not pumping. Under these circumstances so much blood can pool in the legs that the reduced blood return to the heart lowers the blood pressure in the head and the soldier faints.

The problem of returning blood to a giraffe's heart is even more acute because its legs are so much longer. The tightly fitting skin covering the legs appears to reduce the extent of blood pooling. So too does their ability to restrict the flow of blood to the capillaries of the legs. The pumping action of the leg muscles may be a significant factor too, but the whole problem of blood circulation in the giraffe needs more study.

Finally, another problem that a long neck poses is that a lot of air resides in the trachea, the tube that connects the back of the throat with the lungs. This air is unavailable for respiration and so the space it occupies is dead space. The dead space in a giraffe has a volume of about 5 pints (2.5 liters), which has to be moved each time the animal breathes. Consequently, the breathing rate has to increase to compensate for the reduced air flow. A resting giraffe takes about twenty breaths per minute, compared with a human's twelve and an elephant's ten—a very high respiratory rate for such a large animal.

Having seen some of the implications of gigantism in living animals, we can look at the largest animals ever to walk the earth—the dinosaurs. I would give a great deal to be a time traveler. To watch pterosaurs flying or ichthyo-

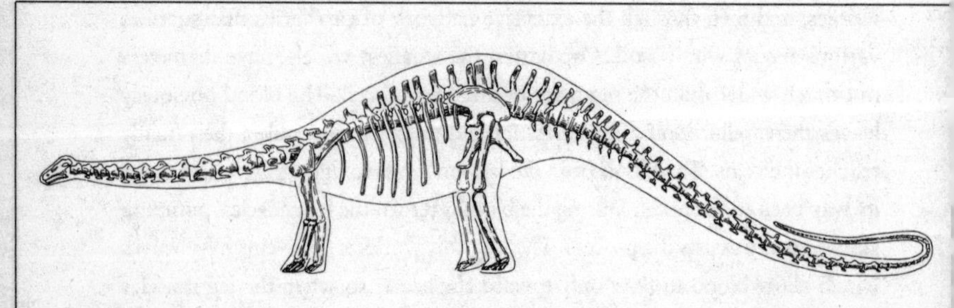

Diplodocus, *Upper Jurassic, western United states; about 85 feet (26 meters) long.*

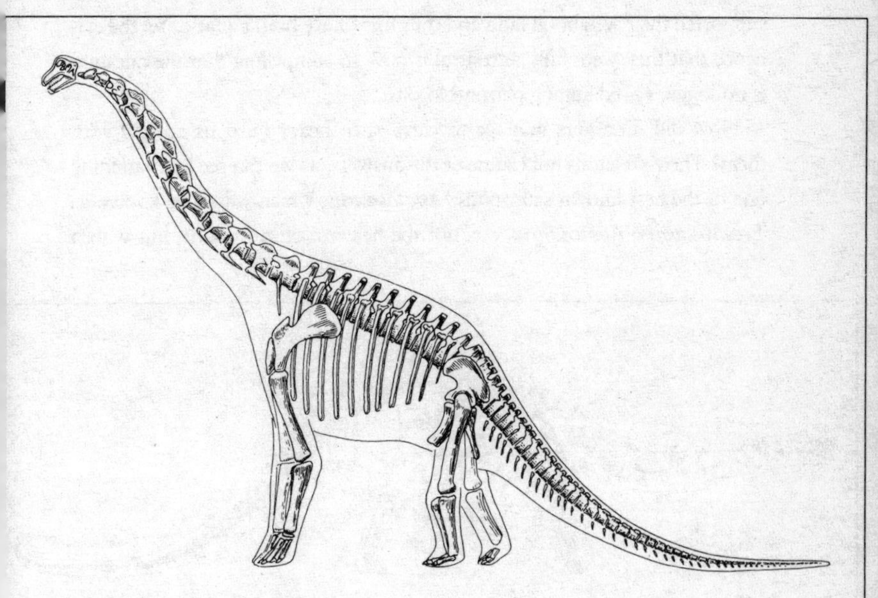

Brachiosaurus, *Upper Jurassic, western United States and east Africa; about 74 feet (22.5 meters) long.*

saurs swimming would be an incomparable joy to me. But if I could witness only one of those denizens of the Mesozoic world, I would have to choose to see a sauropod. The sauropods were the largest of the dinosaurs: *Diplodocus*, *Brachiosaurus*, *Apatosaurus*—names familiar to children everywhere. They were the quintessence of all things dinosaurian and can only be described in superlatives: the longest, the tallest, the heaviest. Imagine standing beside an animal like *Brachiosaurus* that weighed as much as a Boeing 727 airliner. Your head would barely be on a level with its elbows, and you could walk beneath it with your arms above your head and not touch its belly. They were so much larger than anything we know today—*Brachiosaurus*, for example, had an estimated weight of 78 tons, the equivalent of thirteen African elephants—that for a long time paleontologists believed they could not have supported their weight on land and thus must have been aquatic. Yet the evidence that they were fully terrestrial is now so compelling that the question is no longer raised among paleontologists.

How did dinosaurs manage to carry such heavy burdens around with them? Their skeletons hold some of the answers, as we can see by examining one of the best known sauropods, *Apatosaurus*, which was once known as *Brontosaurus*. *Apatosaurus* was not the heaviest of sauropods, but with a

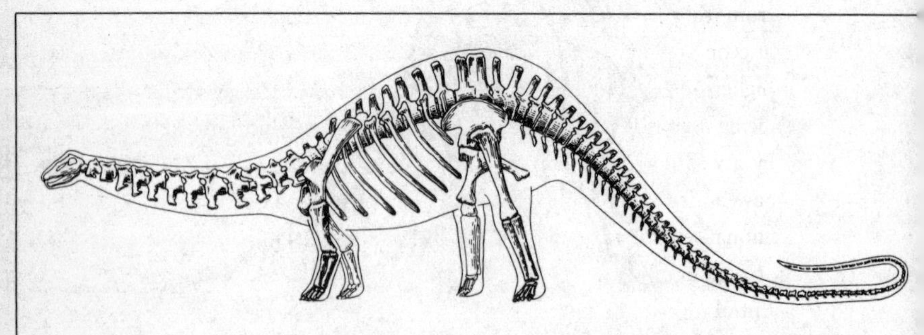

Apatosaurus, *Upper Jurassic, western and southwestern United States; about 65 feet (20 meters) long.*

length of about 65 feet (20 meters) and an estimated weight of some 28 tons, it was about five times heavier than an African elephant. How could its skeleton possibly support such a weight? The vertebral column appears to have been arched between the two limb girdles, a shape maintained partly by individual vertebrae, partly by ligaments, and partly by muscles lying below the vertebrae's transverse processes and muscles in the abdominal region. The neural spines are tall, especially in the dorsal vertebrae between the two girdles. Caudal, or tail vertebrae, especially anterior ones, similarly have tall neural spines. These vertebrae also have a second spine, projecting ventrally, called *chevron bones*, which are found in most reptiles. Having such long processes provided a long leverage and a large attachment area for ligaments and muscles that moved and maintained the vertebral column's shape. Vertebrae in the pelvic region are especially robust. Five of them fused together to form the sacrum, to which the pelvic girdle attached.

The individual vertebrae are massive, especially the dorsal ones that have centra about 1 foot (35 centimeters) in diameter. In life these probably were loaded in bending as beams with compressive stresses dorsally, tensile stresses ventrally, and no stresses in the middle. The vertebral centra did not need to be heavily ossified—that is, to have much bony material—in the middle. Consequently, as a weight-reducing strategy, deep excavations on either side of the centrum (called *pleurocoels*) reduce the centrum to an engineer's I-beam. A similar economy of material, where bone is concentrated in areas subjected to the highest stresses, is seen elsewhere in the vertebrae, bony struts and flanges being used instead of solid bone. Most dorsal vertebrae articulate together by a well developed ball-and-socket joint between their centra. This would have added to the joint's stability and hence its load-bearing capacity. An additional set of articulating surfaces below the level of the anterior and posterior zygapophyses would have strengthened the vertebral column in the dorso-ventral (up and down) plane.

The sauropod's long neck and tail would have required an enormous effort

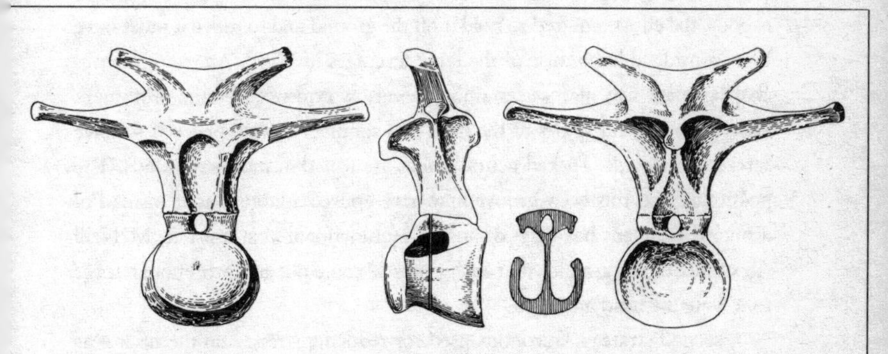

A dorsal vertebra of the sauropod dinosaur Camarasaurus, *seen from the front, left side, and back. A cross-section through the centrum reveals an I-beam structure, a weight-reducing strategy.*

to hold them off the ground, primarily because of the great leverages involved. Dragging the tail may have alleviated some of the problem, though we have reason to believe that the tail may have been kept clear of the ground. Regardless of how the tail was carried, the head was certainly kept elevated and could move up and down and from side to side. But the neck may not have been as flexible as we had thought.

Little attention has been given to how much mobility the vertebral column had, mainly for practical reasons because most continuous series of sauropod vertebrae are permanently joined together in the mounted skeletons displayed in museums. However, an English paleontologist named John Martin was assembling part of a skeleton of *Cetiosaurus*, and had the rare opportunity to examine in detail a well-preserved neck. He discovered that neck movements were severely restricted in the vertical plane because of the way the vertebrae articulated with one another. The head appeared capable of reaching the ground but the neck could not be raised more than 30 degrees from the horizontal. A sound reason for restricting the head's vertical movements is that this would have limited fluctuations in blood pressure experienced by the brain.

Even though head movements may have been limited in some or all sauropods, the effort required to hold it off the ground and to move it must have been considerable because of the large leverages involved. An extensive nuchal ligament was almost certainly present, as evidenced by the prominent forks in the neural spines of the neck and shoulder region, believed to have served as its guide. Forked neural spines are found in most sauropods. The possibility that this bony groove may have housed a large muscle instead of a nuchal ligament has been dismissed by functional anatomist R. McNeill Alexander on the grounds that such a muscle could not generate enough force to elevate the head and neck.

A second strategy sauropods used for reducing stresses on the neck was to have a small and lightly built skull. Not only are the individual bones thin, but there are large perforations in them. Their few teeth are fairly long; some are rod-like, as in *Diplodocus*, and others are more spatulate, as in *Brachiosaurus*. The teeth extend well back in the jaws in some sauropods, like *Bra-*

chiosaurus, but in many others, including *Diplodocus*, they are confined to the front of the skull. The strategy of the sauropod was to have a small but mobile head that gathered food but did not process it. The processing function was almost certainly performed by the gut, probably in a muscular gizzard containing stones for grinding up food, like that of a bird's.

So much for the head and neck, but what about the tail? Was it indeed kept clear of the ground? Like so many other paleontological questions, this one will probably go unanswered because of the lack of data. It appears reasonable to suppose that postures varied from one species to another, possibly even among individuals of the same species. We do have reason to suspect that the tail may have been kept clear of the ground, at least in some sauropods. David Norman, in his outstanding encyclopedia of dinosaurs, suggested that the nuchal ligament may have been continued caudally along the tops of the neural spines of the dorsal vertebrae and along those of the caudal vertebrae. This arrangement would have connected the neck with the tail in such a way that they counterbalanced one another, which makes good mechanical sense. Norman thought that this counterbalancing might account for the rarity of tail-drag marks in sauropod trackways. Negative evidence is unsatisfactory, though.

Positive evidence was provided by the occurrence of the osteological condition called DISH (diffuse idiopathic skeletal hyperostosis) in which articulating bones become fused together by bony outgrowths spanning the joint between them. This condition appears to be non-pathological (not caused by disease or injury) in response to localized regions of high stress. DISH was detected in the caudal regions of half the specimens of *Apatosaurus* and *Diplodocus* that were examined. The area of fusion, involving between two and four contiguous vertebrae, is at about the same level in all skeletons, between the twentieth and twenty-third caudal vertebrae. These fused vertebrae occur at about the level where a gently curving tail would come close to the ground; fusion at this point would have kept a significant portion of the tail clear of the ground. The evidence supports the interpretation that at least some sauropods kept most, if not all, of their tail off the ground.

The weight supported by the vertebral column was transmitted to the legs by the pectoral and pelvic girdles, both of which are characteristically robust, even in the relatively lightly built sauropod *Diplodocus*. In most sauropods (*Brachiosaurus* and its allies excepted) the hind legs are longer than the front ones and the pelvic girdle is correspondingly larger than the pectoral girdle. The scapula is a huge bone and provided a large attachment area for the muscles and ligaments that strapped it to the rib cage. The coracoid is a rounded bone that, together with the scapula, forms the pectoral girdle. It is often fused with the scapula and is similarly large. The three-part pelvic girdle is a massive structure, with the ilium forming a huge thick blade of bone rigidly attached to the sacrum. The bones of both legs are straight and fairly robust, much like those of the elephant, and the terminal positions of their articular surfaces show that they remained essentially vertical. They were therefore loaded as columns, as would be expected, and the cross-sectional supporting area was maximized by being of solid bone, with no marrow cavity. Sauropod feet are very similar to those of the elephant. Although they appear to have walked on the tips of their toes, we can be quite certain that there was an extensive pad of connective tissue beneath the palm and sole, as in elephants. This assumption is supported by the pot-hole shape of their footprints.

The forces required of the skeletal muscles to move such a heavy body are difficult to imagine, and this requirement ultimately translates into food requirements. How much food would a mature sauropod eat in a day? If sauropods were like elephants and were warm-blooded, we could predict their food requirements by extrapolating from the elephant, because of the general relation between metabolic rate and body weight. But could such large animals manage to gather enough food in a day to satisfy a warm-blooded appetite, especially with their relatively small heads?

Robert Bakker, whose radical ideas on dinosaurian energetics and lifestyles have probably stimulated more research on dinosaurs than those of any other paleontologist, considers that all dinosaurs had high metabolic rates. From this has emerged a picture of sauropods as active, warm-blooded animals that were capable not only of running but also of rearing up on their hind legs to browse on tall trees and to defend themselves against predators.

Not all paleontologists, myself included, share this view, but these ideas have had such far-reaching effects that they have been the subject of serious biomechanical discussions and have influenced the restorations of sauropods by artists and recent filmmakers. How likely is this picture of sauropods?

We know that large animals, with their low metabolic rates, eat less food relative to their size than do smaller ones. For contemporary mammalian herbivores, food requirements increase according to the body mass raised to the power of $2/3$. If sauropods had mammalian levels of metabolism, their food requirements would approximate those of scaled-up elephants. *Diplodocus* weighs about three times more than an African elephant, *Apatosaurus* five times, and *Brachiosaurus* thirteen times, so we could estimate their food requirements by raising these numbers by the power of $2/3$. Hence, *Diplodocus* would have to consume $3^{2/3} = 2.1$ times as much as an elephant, *Apatosaurus* almost 3 times as much, and *Brachiosaurus* 5½ times as much. An African elephant spends 75 to 80 percent of every twenty-four hours feeding, so in terms of its food requirements it is close to the maximum size permissible for a mammalian herbivore living in Africa today. Observations made in Sri Lanka on the smaller Asian elephant show that it, too, spends most of its time feeding, so a preoccupation with food is not unique to the African environment.

Although the elephant is the largest land mammal alive today, it is not the largest one that has lived. *Indricotherium* (also known as *Baluchitherium*) is an extinct rhinoceros that lived 35 million years ago during the Oligocene Period, and reached a shoulder height of about 18 feet (5.5 meters). Earlier estimates of this mammalian giant's weight had been set at 26 tons, but more recent investigations revealed that it was more slenderly built and probably weighed only about 11 tons—though the largest individuals may have reached 15 to 20 tons. Even with the lowest estimates, this is still a hefty animal—twice an elephant's weight. How could a large mammal like this obtain enough food in a day when the African elephant has to spend eighteen out of every twenty-four hours feeding? The solution to this problem probably has to do with gut efficiency and the quality of food being eaten. The elephant's digestive tract processes food quickly and what passes out the other end still contains a lot of undigested food. Other mammalian herbivores have more efficient systems and derive more nutrients from their food. Perhaps *Indricotherium* had a more efficient gut than an elephant's, or ate better-quality food, but we shall never know. All we do know is that it did exist and must therefore have satisfied its nutritional requirements.

Given that an 11- to 20-ton *Indricotherium* could eat enough food to satisfy a mammalian level of metabolism, it is conceivable that an 18-ton *Diplodocus* and perhaps a 28-ton *Apatosaurus* could have done the same—but not a 78-ton *Brachiosaurus*. This would only be true, however, if sauropods had guts as efficient as that of *Indricotherium*. We have no way of knowing whether this was so. If sauropods had a digestive efficiency as low as that of the elephant, then it is highly unlikely that even *Diplodocus* could have foraged enough food to support a mammalian level of metabolism—not unless sauropods enjoyed a richer diet. The African elephant takes in rather low-quality food, especially during the dry season; so if sauropods could have found higher-quality food, they might have reduced their consumption relative to that of elephants. Even so, their enormous sizes and their apparent gregarious behavior make it likely that they would have placed tremendous demands on the local flora. Therefore, they may not have been able to consume significantly higher quantities of high-quality food than do elephants—imagine how much food an individual *Brachiosaurus* would have needed to equal the daily food intake of more than five elephants. And *Brachiosaurus* was not the largest of sauropods. *Ultrasaurus*, found in Colorado in 1979, had an estimated length of more than 98 feet (30 meters or over 30 yards!), dwarfing the 74-foot (22.5-meter) *Brachiosaurus*. This is a length increase of 1.32, so the weight increase would be by a factor of $(1.32)^3 = 2.3$—a body mass of almost 180 tons! Such an enormous animal would have had an appetite equivalent to almost ten elephants, assuming a similar digestive efficiency.

To summarize the evidence for food requirements, it is conceivable that the smaller sauropods *could* have had high metabolic rates like those of birds and mammals, but only if their digestive systems had been more efficient than the elephant's. Even if this had been so, it is inconceivable that the giants among them could have obtained enough food to maintain such a high metabolic level.

Other physiological consequences of the sauropods' large sizes would have severely restricted the scope of their activities. Because of its long neck the giraffe's heart has to discharge blood at pressures of about 200 mm Hg to reach the brain. Sauropods had even longer necks, and the vertical distance between their heart and head was almost 7 feet (2 meters) in *Apatosaurus* and 22 feet (6.5 meters) in *Brachiosaurus*. Making allowances for the additional pressures required to force blood through the capillaries of the brain and other organs of the head, the hearts of these two sauropods would have needed to discharge blood at pressures of 216 mm Hg and 568 mm Hg respectively. If sauropods had stood on their hind legs to reach up even higher into the trees, as Bakker has proposed (though not for *Brachiosaurus*), blood pressures in excess of these would have been required. Bakker dismissed the blood pressure problem and suggested that sauropods could have used contractions in the neck muscles to assist the pumping action of the heart. Such muscle contractions, he noted, are probably important in living animals for returning venous blood to the heart. But this argument is not convincing because the pressures involved in returning blood to the heart are so low. For example, the pressure of the blood in a human's superior vena cava, the main vein returning blood to the heart, is between 0 and −2 mm of Hg (the negative pressure is because of the "sucking" action of the heart). The blood pressure in our legs' veins, which are massaged by the leg muscles, is not much higher; it is obvious then, that the pumping action of skeletal muscles has a limited effect on blood pressure.

It has been suggested that sauropods possessed ancillary hearts in their necks to overcome the problem of pumping blood all the way up to their heads. The notion, which has no precedent among the vertebrates, strikes me as being fanciful to a fault. A far more serious proposal is that the blood flowing down the neck in the veins helps draw blood up the arteries to the head, something like a siphon. This idea, championed by physiologists James Hicks and Henry Badeer from Creighton University, Nebraska, deserves special attention because it is likely to be hailed by some paleontologists as the solution to the sauropod's blood pressure problem.

The easiest way to visualize the mechanism is with a simple model made of rigid pipes, like the one the investigators used to illustrate their argument. A pump forces water up a vertical pipe that is connected to a descending one, the circuit being completed at the bottom. Gravity works against the water in the ascending pipe, but this is balanced by the potential energy of the weight of water in the other one. This energy helps draw more water up the ascending pipe, thereby reducing the work of the pump. Hicks and Badeer warn us against referring to this mechanism as a siphon on the grounds that siphons work only in open systems. But this is not so. All that is necessary for the siphon mechanism to work—aside from rigid pipes—is for the hydrostatic pressures in the ascending and descending arms to match.

Once the water is circulating, the only work the pump has to do is to overcome the resistance to the fluid's flow caused by the walls of the pipes (the resistance is called drag). The work of the pump is therefore independent of the height to which the water has to be pumped, and would be the same if the entire model were laid flat. An animal's vascular system differs from this simple model in many regards, including the fact that blood vessels have flexible walls that can collapse. Arteries are exposed to high blood pressures and have thick elastic walls that retain their shape like a rubber hose. Veins, in contrast, have thin and relatively inelastic walls. As veins are exposed to low blood pressures, sometimes lower than the surrounding tissue pressure, they often collapse. When large veins, like the jugular vein in the neck, are partially collapsed, they become dumbbell-shaped in cross section. In an attempt to simulate the living situation, Hicks and Badeer built a working model, using a 1.5-meter (5-foot) long rubber tube to represent the main neck artery (the carotid), and a pair of narrow, rigid-walled plastic tubes to simulate a partially collapsed jugular vein. At the start of the experiment, when the pump had just raised the water to the top of the rubber tube, the pressure at the bottom was just over 110 mm Hg (110 mm Hg being equivalent to a 1.5-

meter column of water). However, once the water had filled the plastic tubes and began pouring out of the bottom, the pressure at the bottom of the rubber tube fell to only 14 mm Hg. This small pressure represents the drag force acting on the fluid. The investigators then raised the ends of the plastic tubes ("jugular vein") until they were horizontal. This eliminated the siphon effect, and the pressure at the bottom of the rubber tube soared to 124 mm Hg (110 + 14 mm Hg). They concluded from these experiments that the heart of a long-necked animal like a giraffe does not have to work to overcome gravity, only to overcome the resistance to the flow of blood by the blood vessels.

Corroborating evidence was seemingly provided by an earlier investigation on the blood pressure of a recumbent giraffe. When the giraffe's head was flat on the ground the pressure of the blood leaving the heart was 170 mm Hg. When the head was raised 1.5 meters (5 feet) above the body, the pressure increased by only 40 mm Hg. Hicks and Badeer argued that if the heart had to compensate for the increased gravitational effect of the blood in the raised neck, the pressure should have increased by 110 mm Hg. Even so, a blood pressure of 210 mm Hg (170 + 40 mm Hg) is still very high, and the investigators attributed this to the high resistance offered to the blood by the thick-walled arteries in the neck, and by the collapsed veins and smaller vessels.

Does the siphon mechanism really work? Roger Seymour, a physiologist at the University of Adelaide, working in collaboration with Alan Hargens, a NASA physiologist, and Timothy Pedley, an applied mathematician from England, showed that it does not. The first point they made was that if the blood left a giraffe's heart at a pressure similar to that of most other mammals (about 100 mm Hg), the pressure would barely be sufficient to drive it up the neck to the head. And once the blood reached the head its pressure would be insufficient to drive it through the capillary bed, which needs a pressure of about 100 mm Hg. They illustrated their point by reference to snakes. Aquatic snakes, like other animals living in water, are essentially unaffected by gravity so their hearts do not have to generate high blood pressures. Consequently, when a snake is taken out of the water and tilted so that its head lies above its heart, insufficient blood reaches the brain and the snake faints.

Arboreal snakes, in contrast, have blood pressures two to three times higher and maintain an adequate blood supply to the brain even when they are vertical. If the siphon mechanism worked in living animals, the aquatic snakes would not faint when held vertically. Furthermore, the distance between their head and heart would be immaterial, but this is not the case. In arboreal snakes the heart is less than a quarter of the way down the length of the body, whereas in aquatic species it is closer to the middle. The rationale for this is that when an arboreal snake is vertical its heart does not have to pump blood as high to reach the head.

The pressure of the water entering the pump in the closed-circuit model is similar to that leaving it, because of the siphon mechanism. If this mechanism worked in animals we would similarly expect the pressure of the blood entering the heart in the veins to be similar to that leaving the heart in the arteries. However, the two pressures are essentially independent. The reason why the pressure energy on the venous side of the heart cannot be transferred to the arterial side has to do with the lungs and the fact that the heart is a double pump. When blood enters the heart it first has to be pumped to the lungs. Passage through their extensive capillary beds drastically reduces its pressure, and on its return to the heart the blood's pressure has to be raised again by the action of the heart muscle. The blood is then discharged to all parts of the body.

To return to the argument about the recumbent giraffe. If the blood pressure did increase by 110 mm Hg when the head was raised, rather than by the observed amount of 40 mm Hg, the blood would reach the head at a pressure of 170 mm Hg. This is about 70 mm Hg higher than it is in a standing giraffe. Increasing the pressure by only 40 mm Hg, however, gives an appropriate pressure of 100 mm Hg at the head. Far from providing supporting evidence for the siphon mechanism, these giraffe observations merely show that the blood pressure to the head drops when giraffes suddenly raise their heads high, just as it does in other mammals.

These arguments against the siphon mechanism are compelling, but before leaving the matter I want to say something about Hicks and Badeer's working model. The heart was represented by a simple one-way pump, there were no capillary beds between the arterial and venous circulations, and the jugular vein was modelled as a pair of non-collapsible tubes. Their model therefore had rigid "veins," not collapsible ones as found in nature. When Seymour and his colleagues repeated the experiment, they substituted a thin-walled collapsible tube for the rigid-walled plastic ones. The tube collapsed: there was no siphon mechanism and therefore no pressure reduction to reduce the work of the pump. Paleontologists should approach the siphon mechanism with caution—all that glitters is not gold!

Whales, the largest of all living animals, have the largest hearts, and that of a 50-ton individual weighs about 440 pounds (200 kilograms). Since the whale is aquatic, its vascular system is not subject to gravity forces, and so it does not have to generate high blood pressures. Vertebrate hearts respond to pressure demands by developing thicker walls. The heart of a 50-ton sauropod has been estimated to be about eight times heavier than that of a 50-ton whale and to be capable of withstanding pressures in excess of 500 mm Hg. A heart that weighs 1.6 tons is hard to imagine. Regardless of whether these estimates and assumptions are valid, they underscore the magnitude of the vascular problems of large land animals with long necks.

Yet another implication of a long neck is the voluminous dead space. The giraffe's 10-foot (3-meter) neck has a dead space of about 5 pints (2.5 liters); so, assuming a trachea of similar diameter, a *Brachiosaurus*'s 30-foot (9.2-meter) neck would have had three times as much dead space, or 15 pints (7.5 liters). This large dead space would have obliged them to breathe at a disproportionately high rate for their body size, as in the giraffe.

A serious problem would have faced sauropods with high metabolic rates: how to cool themselves. The low area-to-volume ratio of elephants makes it difficult for them to keep cool during the hottest part of the day and so they are prone to heat stroke. The problem would be even more acute for sauropods. Even though their long necks and tails could have served as sites for dumping excess heat, it is still difficult to imagine how they could have coped. So why should they go to all the trouble and expense of running high metabolic rates when they could have achieved all the advantages of a constant body temperature at low metabolic rates, simply by virtue of their thermal inertia? Low metabolic rates would also have reduced the problem of getting rid of excess heat and would have minimized food costs.

What does all this mean? The idea that sauropods reared up on their hind legs, raising their heads even higher, can be dismissed from the standpoint of blood pressure alone. Even holding their heads as high as the positions depicted in most mounted skeletons would have presented serious enough difficulties; perhaps they never raised their head much higher than the rest of their body. Such a strategy would have considerably reduced the demands on the heart. In any event sauropods, like giraffes, would have avoided raising and lowering their heads rapidly. We can therefore dismiss the idea, popular in some restorations, that sauropods threw themselves into combat with predators. Gone, too, is the notion that sauropods could trot and gambol along as depicted in so many illustrations.

Gigantism is not the prerogative of dinosaurs, and many other animal groups have evolved toward gigantism. Indeed, the trend is so common in the fossil record that the phenomenon has been described as a rule, named Cope's Rule in honor of Edward Drinker Cope, who first looked into it. Like most generalizations exceptions abound; some of the earliest ichthyosaurs, for example, are the largest ones, and the last of the group is only of modest size. Yet it is generally true that most groups of large animals have evolved from small ancestors, and many lineages of animals did evolve toward large size. The trend must have selective advantages. Aside from discouraging predators, giants need less food per unit of body weight than do smaller animals; unlike large cars, they are more fuel efficient. But large animals produce far fewer offspring than do smaller ones, and they have longer generation times. Hence their population densities are much smaller—there are always more mice in a field than there are deer. Giants therefore make a trade-off between relative immunity from predators and fuel economy, on the one hand, and lower reproductive rates, therefore lower population densities, on the other. When times are good all is well, but when disaster strikes giants are vulnerable. A drought in Africa can decimate local elephant populations, but enough small animals survive to continue their race.

If we could travel back in time and catch a glimpse of a herd of large sauropods, what might we see? Once we had recovered our composure at the sight of such large and magnificent creatures, we would probably be struck by their slow pace of life. Some might be standing motionless, surveying the world around them with expressionless eyes. Others, necks gracefully curved to the ground, might be feeding, using the advantage of their long reach to sweep a curving swath on either side of them. They would probably look much thinner than usually depicted by artists—gazelles rather than wildebeest. Some might be ambling along at an unhurried pace. But I would be wary about being fooled by their apparent lack of speed; their legs are twice as long as I am tall and I would probably have to run hard to keep up with them.

Suppose that a predatory dinosaur appeared on the scene. It might weigh over a ton, but would be dwarfed beside the gargantuans. One of the sauropods closest to the carnivore might stop feeding, slowly raise its head, and peer sideways at the intruder. Sauropods were not great thinkers, but this one would probably recognize that its superior size gave it immunity from attack. After a moment's contemplation it might lower its head and resume feeding. Any young sauropods in the vicinity might be vulnerable to attack—good reason for keeping close to their parents.

We do not know how long the adults remained fertile, but it is possible that they continued breeding throughout life, as modern tortoises appear to do—remember the link between longevity and long reproductive life. If this were the case, some of the largest sauropods might have lived for two centuries or more.

I have implied that sauropods were not well endowed with gray matter, and this is based on the small size of their cranium. However, as we will see in the next chapter, the assessment of the intellectual abilities of animals is a far more complex issue than the simple matter of brain size.

BRAINS: FROM THE MASSIVE TO THE MINUTE

When we moved into our new house beside the golf course we were soon befriended by the local squirrels, who took a great deal of interest in our bird-feeder. They had no difficulty shinning up the steel pipe supporting the well-stocked granary, and although we tried to discourage them by leaving piles of seeds and nuts on the ground, they still preferred to raid the feeder. But we did not mind, and it used to amuse us to see legions of squirrels bounding across the golf course to converge on the McGowans' for breakfast. Then I planted some rhododendrons—plants that need a lot of tender loving care in our northern clime—and the squirrels trashed them. And so began the squirrel wars.

My first act of hostility—and I felt very hostile after putting so much effort into my rhododendrons—was to launch a resettlement program. I borrowed a small mammal trap, baited it with peanut butter, and waited to take my first prisoner. Squirrels are inquisitive animals, and it was not long before I had my first customer. But they are also very suspicious, and it took an age before a squirrel finally ventured inside, tripping the trap-door. With great jubilation I drove out into the countryside, released the first captive, and dashed back home for more. I re-baited the trap, put it outdoors, and waited. The squirrels came, they saw, but they would not enter—somehow word had got out that this was a trap. I realized I was dealing with an intelligent enemy.

If I could not take prisoners I would cut off their supply lines, and to this end I visited the local hardware store and purchased a dinner-plate-sized disc of metal designed to prevent squirrels from shinning up poles. They loved it! Far from preventing them from reaching their objective I think the new assault course added to their enjoyment. I countered by replacing the metal disc with a plexiglass one that was twice as large, and was gratified to find that they could not get past it. The perplexed expression on their tiny squirrel faces was a joy to behold, and the marauding rodents had to content themselves with scraps from the birds' table.

My euphoric state lasted for several days; then we awoke one morning to find a squirrel *inside* the bird-feeder! How the raid had been accomplished was a complete mystery to us, and, hoping to catch one of them in the act, we kept a sharp look-out for the next few days. We had no success, but we did see the same individual inside the feeder again. Then we started noticing that other squirrels had learned the trick too, but we still did not know how they were doing it. The mystery was resolved a week later when my wife saw a squirrel climb the nearby fence, launch itself into the air, and land, commando style, on the roof of the bird-feeder. We had to admire their ingenuity—squirrels are very smart—but war is war and we countered by moving the feeder further away from the fence. This did the trick, and the squirrels have been unable to reach the bird seed ever since. But their luck did change for a few weeks last winter when the hard-packed snow was so deep that they could leap up onto the plexiglass disc from the ground.

Cats, in contrast to squirrels, are not very smart. True, they always manage to appropriate the most comfortable chairs in the house, and they are good at manipulating their owners, so I should not be too dismissive of their intellectual abilities, but they do not seem very good at working things out for themselves. Our cat will whine to go out, turn on her tail when she sees that it is raining, and whine to go out again a few minutes later. Even my wife, staunch defender of the cat, admits that squirrels can run intellectual circles around our feline. But why the difference?

A squirrel's brain is about the size of a large grape and weighs a quarter of an ounce (8 grams); a cat's brain is egg-sized and weighs about four times as much. The human brain, for comparison, is about as big as a cantaloupe melon and has an average weight of nearly 3 pounds (1.3 kilograms), and an African elephant's brain, about the size of a watermelon, weighs in at 10 pounds (4.5 kilograms). Intellectual ability clearly has little to do with absolute brain size, so we need some measure of relative brain size that takes account of differences in body size. Simply comparing brain weight with body weight is no help. A cat's brain, for example, accounts for 1.6 percent of its body weight, while a lion's is only about 0.13 percent, even though the lion is not intellectually inferior. The reason for this discrepancy is that brain size and body size are related to each other by a power function, as with metabolic rates. If logarithmic data for brain mass are plotted against body mass for many mammals, a straight-line graph with a slope of 3/4 results. This is the same exponent as for metabolic rate, and has led to all manner of speculations on a linkage between metabolic rate and brain size. We will see later that there is probably no connection between the two, certainly not a universal one.

Harry Jerison, a pioneer in the field of relative brain size, plotted a double logarithmic graph of brain mass against body mass for almost two hundred species of vertebrates. The results were interesting because the points clustered around two separate but parallel straight lines. The upper line corresponded to birds and mammals (the "higher" vertebrates), while the lower line, whose members had brains only one-tenth as large, represented fishes, amphibians, and reptiles (the "lower" vertebrates). Some points lay well above or well below their respective lines, showing that these species had par-

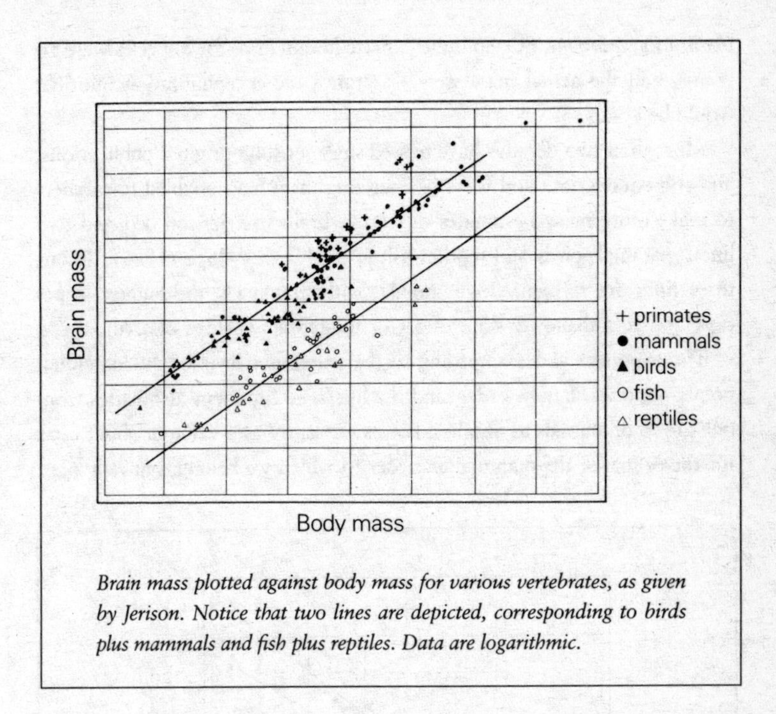

Brain mass plotted against body mass for various vertebrates, as given by Jerison. Notice that two lines are depicted, corresponding to birds plus mammals and fish plus reptiles. Data are logarithmic.

ticularly large or small brains for their body size. Our own species, for example, was well above the bird-mammal line while the blue whale fell well below it.

Here was a way of expressing relative brain size that took account of the general trend for large animals to have correspondingly smaller brains. (Large animals have larger brains than small animals in absolute terms, but, expressed as a percentage of body mass, their brains are relatively smaller.) Jerison calculated the expected brain mass of the species in question; that is, the brain mass if the point lay on the graph. The expected mass could be read directly from the graph, or, preferably, calculated by using the equation for the graph. He then compared expected brain mass with the actual mass, by dividing the latter by the former. He termed the resulting number the *ence-*

phalization quotient, EQ. So if the expected brain mass for a species were 20 grams, and the actual mass were 60 grams, the encephalization quotient would be 60/20 = 3.

More than two decades have passed since Jerison's original publications, and subsequent data and ways of analyzing them have enabled researchers to make more refined estimates of relative brain size. Jerison depicted two lines, mammals-birds and reptiles-fishes, each with a slope of 0.67. Today, three lines are recognized: mammals (with marginally the highest slope: 0.75), birds (a shallower slope of 0.56), and reptiles (a slope of 0.56).

The exponent varies according to the taxonomic level of the individual points—species, family, order, and the like. The exponent also varies from one group of animals to another. For example, it has a value of about 0.92 for the primates, the mammalian order to which we belong, but only 0.46

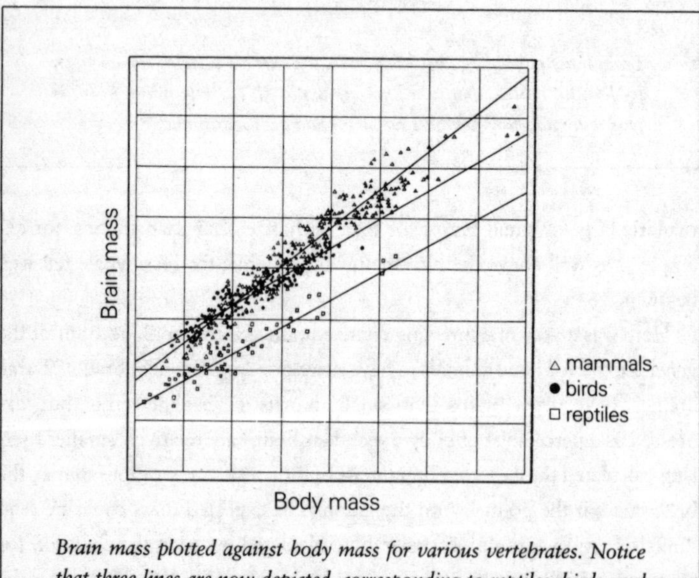

Brain mass plotted against body mass for various vertebrates. Notice that three lines are now depicted, corresponding to reptiles, birds, and mammals. Data are logarithmic.

for the cetaceans (whales and dolphins). Hence the difference in brain size between small and large individuals is much greater in primates than in whales. These examples underscore the tremendous influence of an animal's phylogeny, or evolutionary history, on its anatomy. Because of these new findings Jerison's original encephalization quotients are now seldom used. Instead, the brain of the species of interest, say a gorilla, is compared with the group graph—in this case the primates—to see how it compares in mass.

So far we have seen that, within a given group of animals, small species have larger brains relative to their body mass than large species; for example, sheep have larger brains than cows, on a percent-of-body-weight basis. The same is true for developing individuals. Kittens have relatively larger brains than cats, just as our own babies have relatively larger brains, hence larger heads, than adults. The growth of the brain is not unusual in this regard, and allometric growth is the norm rather than the exception.

The fact that a human adult has a relatively smaller brain than a baby, or that a lion has a relatively smaller brain than a domestic cat, would not be expected to result in any differences in intellectual abilities between large and small; but some differences might exist between, say, the polecat (*Mustela putorius*) and a similarly sized and closely related carnivore, the sable (*Martes zibellina*), whose brain is more than twice as big as a polecat's. We might expect the sable to be smarter, and this might be true, but there is not a simple relation between brain size and intellectual ability. The absence of a simple relationship is clearly evident for our own species, whose adult brain sizes range from about 2.2 pounds (1 kilogram) to 4.4 pounds (2 kilograms), apparently without any correlation with intellect. Indeed, the relation between brain size and intelligence has been hotly debated for well over a century. The great French anatomist Georges Cuvier, who was noted for his large head, was one of the great intellects of his day. When he died his colleagues, bless their sentimental souls, dissected his body and removed his brain. Cuvier's brain tipped the scales at an impressive 4 pounds (1.83 kilograms), which is about 40 percent heavier than the average human brain. It is purported that Lord Byron, Oliver Cromwell, and Jonathan Swift were in the same cerebral league, whereas Albert Einstein's brain was apparently of average size. What is missing from these few data, of course, is body weight, and when you take into account the large size range within our species, it is hardly surprising to find such a wide range in brain sizes. Much of the variation in body weight, though, is determined by diet, and although some of us may have gained in girth since reaching maturity, our brains have remained the same size!

Perhaps a study of *relative* brain size and intellectual ability might reveal a correlation, but this is a far more sensitive issue in our politically correct times than it was during Cuvier's day. The notion that large brains may have something to do with intellect is almost a heresy; even so, at the risk of being so branded, I should mention the findings of psychologist L. Willerman and his colleagues at the University of Texas in Austin. They pointed out that although head size is often used as a measure of brain size, it does not give a very accurate estimate. Brains are sometimes weighed directly, but often after they have been preserved and undergone shrinkage, which introduces errors. To overcome these problems the psychologists measured brain volumes of living volunteers by using magnetic resonance imaging (MRI), a technique that generates a series of thin photographic "slices" through the brain. Twenty male and twenty female volunteers—all first-year university students—were divided into four equal groups according to their IQ and sex. Thus the group was divided into high-IQ women, high-IQ men, average-IQ women, and average-IQ men. The men had larger brains than the women because of their larger body size, but the most interesting finding was that large brains correlated with high IQs in both men and women. This is only one investigation, however, involving a fairly small sample size, so we should not read too much into the results. Indeed, the number of correlations between brain size and mental ability are few and far between—and some are rather puzzling.

Carnivores such as cats that stalk their prey, seizing and manipulating them with their front paws, have relatively large brains. Insectivorous carnivores like badgers, in contrast, have relatively small brains. These differences are not correlated with diet per se, but with the way that prey are captured—more coordination and mental activity are required in stalking and capturing a large animal than in seizing a small insect. In a similar fashion, primates and rodents that feed on well-spaced food items, such as fruit, have relatively larger brains than those that feed on leaves and shoots. This is so because searching for limited food items is more complicated than feeding on leaves, which are readily available and can be found everywhere. A similar argument has been used to explain why fruit-eating bats have relatively larger brains than those that capture insects on the wing. At first glance this seems unreasonable: Surely capturing insects while in flight would place greater demands on a bat's brain and sensory apparatus than foraging for fruit? But fruits have an unpredictable distribution, both temporally and spatially, and bats have to rely on several sense organs and on their memory to find them.

Foraging strategies also correlate with brain size in birds. Birds that hunt from branch to branch have relatively larger brains than those that forage on the ground. This might be because moving in three dimensions, as in flitting between branches, requires more mental capacity than hunting in two dimensions on the ground. This is also the case for squirrels; arboreal ones have relatively larger brains than those that live on the ground. One has only to think of the acrobatic feats of squirrels as they leap between branches to realize that this activity requires a great deal of eye-limb coordination and split-second timing. Little wonder that squirrels have such agile minds. While quick thinking and arboreal life appear to go hand in hand, I could be making a cause-and-effect correlation where none exists. And if I wanted a reminder of the dangers of constructing just-so stories, I need only consider the correlations between brain size and mating activities in birds and mammals. Monogamous and polygamous species occur in both groups. Among mammals, the primates show a correlation between large brains and polygamy. Perhaps this is because having several mates is more complex than having one. But this reasoning is erroneous because in birds large brains are correlated with monogamy, not polygamy. We are obviously getting mixed messages about the correlation between brain size and mating preferences. Nevertheless, birds and mammals have consistent correlations between brain size and the mode of development of their offspring.

Many animals, including cats, dogs, sparrows, and starlings, are born or hatched in a helpless state. These youngsters, which are often naked, blind, and unable to walk or feed themselves, are called *altricial*. In contrast are the *precocial* offspring of animals like chickens, gulls, horses, and dolphins. These youngsters are hatched or born in an advanced state of development, with open eyes and the ability to walk and feed themselves. Being able to fend for themselves requires high levels of brain activity, and therefore it is not surprising that precocial birds and mammals have relatively larger brains at hatching or birth than their altricial relatives. This is because the parents of precocial species invest more energy in their offspring's embryonic development than do altricial species, as reflected in birds laying larger eggs, and in mammals having longer gestation periods. Altricial species, in contrast, lay smaller eggs or have shorter gestation periods; their strategy is to invest more energy after their offspring have hatched or been born. The contrast is most apparent in birds: parents of altricial species dash back and forth in a frenzy to satisfy their demanding offspring, while precocial species take a more casual approach to parenting. After hatching, the small-brained altricial birds soon catch up with their precocial relatives and overtake them. Consequently, most altricial adults have larger brains than do most precocial species. Among the large-brained altricial species are parrots, owls, penguins, and hawks; the smaller-brained precocial species include chickens, game birds, nightjars, ostriches, and emus.

Young altricial mammals similarly catch up with their larger-brained precocial relatives. In contrast to birds, however, they do not overtake them, so there is no difference in relative brain sizes between adults of precocial and altricial species. The high levels of embryonic growth in precocial birds and mammals tend to be compensated for by lower levels of post-embryonic growth. The reverse is true for altricial species. Our own species is exceptional because high levels of pre-natal growth during a long gestation period are followed by high levels of brain growth after birth. The result? An adult whose brain is bigger relative to the body size than in any other animal.

So far I have dwelt on the size of the entire brain, and I have been hard-pressed to find many explanations for why some species have relatively larger brains than others. What about the size of different parts of the brain? Surely we could expect to find correlations between brain regions associated with particular functions and an animal's life-style? One can readily see major differences between, say, a turtle's brain and a dolphin's. Reptiles are not great thinkers, and that part of the brain associated with higher mental activities, the cerebral hemispheres, are small and simple, whereas in mammals they are large and convoluted. Yet when it comes to detecting correlations between development and life-style among similar animals, the differences are far less obvious. Take, for example, the development of the cerebellum, the hind part

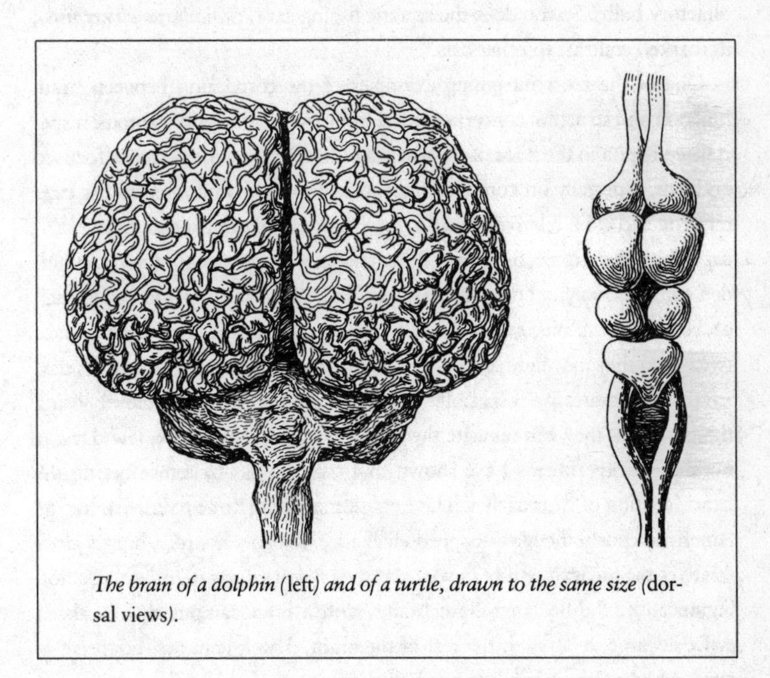

The brain of a dolphin (left) *and of a turtle, drawn to the same size* (dorsal views).

of the brain associated with coordination and posture. Birds that hunt from branch to branch have larger cerebella than those that hunt on the ground, which makes perfectly good sense. But birds that hunt on the wing have *smaller* cerebella, which makes no sense at all. Furthermore, variations in the size of the optic lobes, which are concerned with sight, do not appear to be correlated with whether the birds are active at night or during the day. Nocturnal birds, though, do have larger olfactory bulbs, suggesting that the sense of smell may be important to them when low light levels diminish visual acuity.

The correlation between the sense of smell and olfactory bulb size has been well documented for mammalian carnivores by John Gittleman of the University of Tennessee. Otters, for example, which spend much of their time in water, where olfactory communication is of reduced importance, have small olfactory bulbs. So too does the aquatic fishing cat (*Prionailurus viverrinus*), in marked contrast to other cats.

One of the most intriguing examples of the correlation between brain function and structure concerns spatial memory and the hippocampus, a specialized region in the floor of the forebrain. Much of the research has focused on birds, especially on comparisons between those that store food for later retrieval and close relatives that do not. Black-capped chickadees (*Parus atricapillus*), for example, like many other chickadees and tits, store food in hidden caches throughout their home range. Only one food item is stored in each cache—a seed, a nut, or an insect. They do not use the same hiding place twice, and they may hide several hundred items during a normal winter's day. Even more impressive, especially for people like me who are forever losing things, is that they can relocate their hidden stores, even after a few days. A number of experiments have shown that they do this by remembering the exact location of their caches. The English marsh tit (*Parus palustris*), like its American cousin the black-capped chickadee, is also a hoarder, but its close relatives the great tit (*Parus major*) and the blue tit (*Parus caeruleus*) are not. Significantly, the hoarders have brains with a hippocampus that is about twice as large, relative to the rest of the brain, than it is in non-hoarders of similar body size. A black-capped chickadee (hoarder) weighing 11 grams, for example, has a hippocampus with a volume of 14.26 cubic millimeters compared with a volume of only 11.2 cubic millimeters for a 20-gram great tit (non-hoarder).

Food storage is especially important for birds that spend the winter in northern climates. It has been estimated that Siberian tits (*Parus cinctus*) may store as many as half a million items in a year; they spend most of the daylight hours of fall foraging and storing their seeds in places that will remain free of snow. Among the most specialized hoarders is Clark's nutcracker (*Nucifraga columbiana*), a blue-jay-size bird belonging to the same family as jays and crows. They specialize in pine seeds, which they pry from unripe cones. Carrying the seeds in a throat pouch, they travel up to 22 kilometers away, to south-facing "storing slopes." The slopes are chosen because they remain relatively snow-free, and one nutcracker can cache up to 33,000 seeds in about 6,600 separate locations.

Many rodents store food too; predictably the hippocampus is relatively larger in those that hide food in caches than in those that do not. The bannertail kangaroo rat (*Dipodomys spectabilis*), a small hopping animal that lives in the desert, hoards its seeds inside its burrow and therefore does not need any spatial memory to locate them. Merriam's kangaroo rat (*Dipodomys merriami*), in contrast, hides its seeds in scattered locations and depends upon its spatial memory to relocate them. Predictably, its hippocampus is relatively much larger than in the bannertail kangaroo rat.

In addition to differences in hippocampal development among species of rodents, there are also differences between males and females of the same species, related to their mating patterns. Voles are mouselike rodents found in many parts of the world. Some, like the American pine vole (*Microtus pinetorum*), are monogamous. Since the male confines his amorous attentions to one female, he does not have to go off in search of others, and the size of his range is similar to that of the female. Males of polygamous species, like the American meadow vole (*Microtus pennsylvanicus*), have numerous females and therefore do much traveling in pursuit of mates. Their ranges are consequently four to seven times larger than those of the females, so they have a greater need for spatial awareness. Predictably, the males of polygamous species outperform the females in laboratory experiments that test place-learning and route-learning skills; monogamous males and females perform equally well. These differences are reflected in the size of the hippocampus. Relative to the rest of the brain, the hippocampus is significantly larger in male meadow voles (polygamous) than in females, but there are no differences in the pine vole (monogamous).

Hippocampal development is also associated with homing abilities in pigeons. The ability of birds to navigate over hundreds or even thousands of miles has been extensively studied for many years. The subject is too large for inclusion here; suffice it to say that experiments with homing pigeons have shown that they use geomagnetic, celestial, and olfactory cues to find their way back home. Although damage to the hippocampus does not impair a pigeon's ability to orientate toward home when it is released from far afield, it usually prevents it from returning home. Even when released within sight of its loft, a pigeon with a damaged hippocampus is often unable to reach home because of impairment in its recognition of spatial cues, like the surrounding houses and trees.

Not all breeds of pigeon are able to home; when the brains of homing pigeons are compared with those of non-homing breeds, they are found to have a relatively larger hippocampus. All breeds of pigeon were evolved by artificial selection from the wild rock pigeon, during a relatively short period of a few centuries. This illustrates how rapidly changes in brain development can evolve. Changes can also occur during an individual's growth. For example, laboratory rats raised in complex environments grow up with a relatively larger cortex (part of the forebrain) than rats raised in simple cages.

Oxford University biologist Nicky Clayton, working in collaboration with her colleague John Krebs, has been conducting some exciting research on spatial memory in birds. The research is so new that it had not appeared in print at the time of writing this chapter, but they kindly agreed to let me include their findings here. In an elegant series of experiments, young marsh tits (hoarders) were reared in isolation from other birds. The young birds were given various levels of practice at storing and finding seeds at various stages during their development. These ranged from practices every day to none at all. Those individuals that were allowed to hide and seek seeds grew up to have a well-developed hippocampus, like their kin in the wild. But the deprived birds grew up with a relatively small hippocampus. Neither the amount of practice nor the age at which the exercises began had any effect on the development of the hippocampus. A certain minimum level of practice was required, however, and birds that were allowed to store and retrieve only one seed per day grew up with a small hippocampus. The fact that lack of use of the brain during an individual's development can materially affect its growth is rather startling, and raises some interesting questions for other species, including our own.

As mentioned earlier, the slope for the double logarithmic graph of brain mass plotted against body mass is 0.75 for mammals, the same as for their metabolic rates. This fact has persuaded some specialists to argue that brain size is linked with metabolic rate, and that the size of an animal's brain is determined by the metabolic rate of its mother. This line of reasoning applies only to mammals, not to birds or reptiles, whose brain graph has a slope of about 0.56 rather than 0.75. Moreover, exceptions abound. For example, mammals with high metabolic rates for their body size should produce large-brained offspring, but this is not so. Also, selected groups of mammals have markedly different brain-graph slopes from 0.75. All of this refutes a universal relation between metabolic rate and brain size.

The largest brain in our world, and probably the largest living computer that has ever existed on our planet, is that of the sperm whale. The record, claimed for a 49-foot-long (15-meter) male killed in 1949, weighed just over 20 pounds (9.2 kilograms). The blue whale, the largest of all living animals, reaches twice the length of a sperm whale, but its brain is both relatively and absolutely smaller and does not exceed that of an African elephant. The discrepancy between these two whales occurs because they belong to different groups. The blue whale is a mysticete or baleen whale (sometimes called whalebone whale), while the sperm whale is an odontocete or toothed whale, belonging to the same group as dolphins, porpoises, and the killer whale. The odontocetes are fish-eaters and carnivores, but the mysticetes feed on small crustaceans that they sieve from the water. Mysticetes, being warm-blooded and weighing many tons, have to eat prodigious quantities of oceanic minutiae to satisfy their appetites. To this end they have an enormous mouth that occupies about one-third of their total body length. The mouth is lined with a curtain of bristles—called baleen or whalebone—for straining the food. The head is primarily a filtering plant, and the brain occupies such a small part that it is relatively small compared with the rest of the body. Odontocetes, in contrast, have a regular-size mouth so their head is not unusually large, nor is their brain unusually small. Indeed, probably because they echolocate and live complex social lives, which require well-developed intellectual abilities, they have a relatively large brain. Consequently, odontocetes lie above the brain/body graph, together with primates and bats, while mysticetes lie beneath it. The single exception is the sperm whale, an odontocete

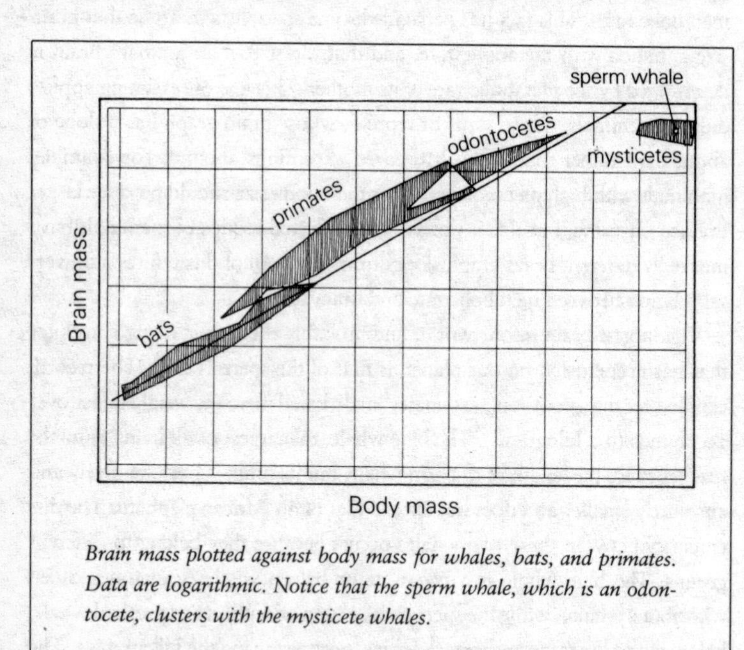

Brain mass plotted against body mass for whales, bats, and primates. Data are logarithmic. Notice that the sperm whale, which is an odontocete, clusters with the mysticete whales.

with a large head and a small brain relative to its body size, though in absolute terms the brain is huge. Unlike other odontocetes it does not echolocate, nor does it have a complex social life. These differences are thought to account for its small brain.

Brain cells are among the most metabolically active cells in the body, and hence the brain demands a lot of oxygen. This explains why severe brain damage often results when the oxygen supply to the brain is temporarily interrupted, as during near drownings, seizures, and mishaps during surgery.

Mysticetes often dive for half an hour and more, and sperm whales can stay down a staggering ninety minutes. How can they supply sufficient oxygen to the brain? Some specialists have suggested that large brains would be a severe handicap to deep divers, arguing that this is why mysticetes and the sperm whale have small brains. This seems a valid argument, but some seals with average-sized brains make long dives, and so the argument does not hold up. The question remains: how do deep-diving mammals overcome the problem of supplying sufficient oxygen to the brain? Several mechanisms come into play. First, when an animal dives, the blood vascular system severely reduces blood flow to other parts of the body, redirecting it to the central nervous system and heart. Diving mammals also have more blood than do terrestrial mammals; this blood is richer in hemoglobin, the red pigment that combines with oxygen to increase the blood's oxygen-carrying capacity. Divers also have elevated levels of myoglobin, a pigment similar to hemoglobin, which is an important oxygen reservoir found in muscles.

Having spent time thinking about brains in the largest living animals, let us turn to the other end of the size spectrum. Salamanders are tailed amphibians like the newts I caught in my youth. They are among the smallest of living vertebrates. Most of them catch flies and other small prey by flicking out their elongated tongues. It is not surprising that they are very dependent on vision, and hence on the parts of the brain associated with sight and coordination. Although some salamanders are quite small, only about 1.5 inches (4 centimeters) long, other species are several inches long. The large size range, from the modest to the minute, attracted the attention of Gerhard Roth at the University of Bremen and his colleagues. They wanted to know whether there were limits to the degree of miniaturization. How small could a salamander become before its brain and eyes no longer were capable of "seeing" properly and steering the tongue to its target? This led to detailed comparisons among seven species of salamanders, ranging in length from 1.5 to 3.5 inches (4 to 9 centimeters). They discovered a wide range of modifications that compensated for the reduction in body size.

One of the most noticeable changes is a modification in the shape of the brain. Smaller species have narrow brains because of the space requirement of the eyes, which are relatively larger as head size diminishes. If the eyes were smaller in accordance with the smaller head, they would become too small to be functional. The smallest salamanders have eyes sufficiently sensitive to detect small prey—fruit flies and the like—at distances of 12 to 18 inches (30 to 50 centimeters). Large eyes in a small head makes perfectly sound functional sense for these small salamanders, but bear in mind that this particular allometric relationship, where small animals have large eyes, is almost universal. In addition to having large eyes, the smaller species also have more light-sensitive cells in the retina, partly due to a denser packing of the retinal cells. Vertebrate brains have hollow central spaces, called ventricles, and these

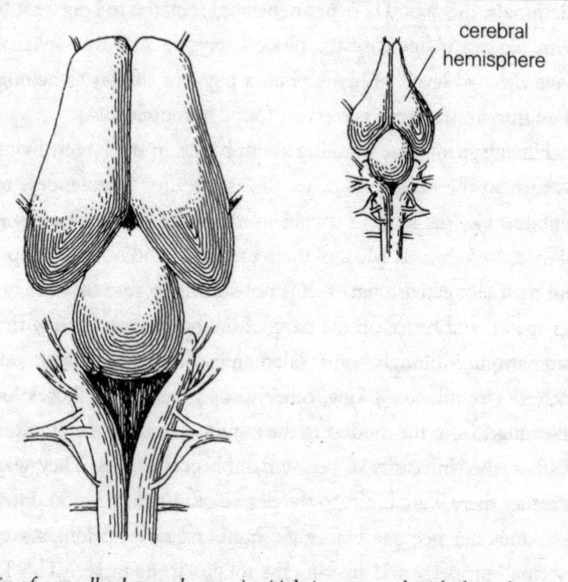

The brain of a small salamander species (right) *compared with that of a larger species, drawn to the same scale. Notice that the cerebral hemispheres of the smaller species are much narrower, due to the encroachment of the eyes. (The species are* Thorius narisovalis *[*right*] and* Bolitoglossa subpalmata *[*left*]).*

are smaller in the smaller species. The brain is also relatively larger in the small salamanders. The consequence of these two modifications is that, although the smallest species' head is only one-twenty-seventh that of the largest one, its brain is only one-eighth as large. Other modifications include a higher density of brain cells, more gray matter than white matter, and somewhat smaller nerve cells, though I should caution that some of these differences may have more to do with the evolutionary relationships among the species than with miniaturization. The net result is that the smallest species, whose head is one-twenty-seventh of the largest one, has one-third fewer nerve cells in the part of the brain concerned with vision. This is an outstanding achievement of miniaturization and packaging.

If you think small salamanders have tiny brains, consider the minuscule brains of their insect prey. Like other vertebrates, our bodies are under the brain's central control; if the brain is severely damaged or destroyed, life is terminated. Control of an insect's body, in contrast, is largely decentralized and mostly under the local control of nodules, called nerve ganglia, that function as mini-brains. These ganglia occur in pairs throughout the body's length. Each ganglion comprises a small clump of nerve cells. Because of the diminished emphasis on central control, an insect's brain can be destroyed yet most of its body will continue functioning, almost the same as before. Each leg, for example, is controlled by its own ganglion and can still be trained to respond to stimuli even after the brain has been removed. Nevertheless, the brain performs a vital role in coordinating the body's local actions, so if an insect's brain is destroyed it may try to feed, walk, and fly all at the same time. The brain is also the center for an insect's behavioral repertoire, which can be quite complex. Consider, for instance, the complicated caste system of honeybees; or the way they build hexagonal cells of wax where they store their honey; or the dance of the bee, in which foraging workers return to the colony and communicate where they have been successful in finding pollen. All of this is done with a brain that is just over a millimeter across—the size of a pinhead.

Ants come in a great variety of sizes and share certain behavioral activities that are typical of the group as a whole. These activities include the licking and carrying of eggs, larvae, and pupae; the feeding of larvae; self-grooming and the grooming of others; and the removal of waste from the nest. Some ants have far larger behavioral repertoires than others, and this is correlated with body size. Not surprisingly, large ants have absolutely larger heads than small ones, hence larger brains with more neurons. This appears to explain why large ants can perform more varied activities than small ones.

Many insects can generate sounds; suitors often use these sounds during courtship to attract mates. Some flying insects have even evolved a sophisticated sound-generating system as a defense against insectivorous bats. Bats locate their prey by echolocation, that is, by emitting bursts of high-frequency sound that bounce off potential prey and reveal their bearing and distance. Certain insects can emit similar bursts of sound that interfere with the bats' sonar, essentially jamming it. These insects have brains no bigger than that of the bee. But there are some flying insects, tiny wasps belonging to the genus *Trichogramma*, whose entire bodies are about this size. These minute (0.6 to 0.8 millimeters) wasps have been the subject of considerable scientific interest, mainly because of their egg-laying habits. They lay their eggs inside those of other insects, and when the wasp larvae hatch they devour the contents of the surrounding egg, thereby killing it. One of their hosts, the spruce budworm, is an economically important pest of the forest industry; hence foresters are interested in the prospects of using *Trichogramma* as a biological control. Before laying her eggs the female wasp inspects the host egg to determine how many eggs to inject inside it. If it is a large egg she will lay several of her own, but if it is small she might lay only one. The wasp's strategy is to lay as many eggs as the host's egg can support. If she lays too many eggs her hatching larvae will starve, but if she lays too few eggs she has wasted an opportunity to rear more offspring.

The importance of allocating the right number of eggs to a host egg has been recognized for many years. But it was not understood how the female wasp arrived at her decision until Jonathan Schmidt of the University of Guelph and Berry Smith of the University of Toronto carried out their innovative laboratory experiments and painstaking observations. Before laying her eggs the wasp walks over the prospective host's egg, drumming the surface with her antennae as she goes. The duration of the walk, from when she leaves the ground until she reaches the ground on the other side, enables her to estimate the volume of the egg: the larger the egg, the longer it takes her to complete her walk. Sometimes an egg will not be lying flat on the ground but will be raised, as on a stalk. In such cases her walk would be considerably longer than normal, causing her to overestimate its size. The wasp can somehow tell when this has happened, though, and instead of estimating the volume by timing the walk, she deduces the volume by calculating the curvature of the surface. This calculation appears to involve measuring the angles between the segments of her antennae while she touches the egg's surface. The wasp also decides the ratio of male and female eggs that she deposits (the fertilized eggs that develop into females are kept separate from the unfertilized ones that develop into males). It is astonishing that all of these computations are made with a brain containing about 10,000 nerve cells that is less than half the size of the period at the end of this sentence. When I confessed to Dr. Schmidt that I could not understand how this minuscule brain could carry out such complex calculations, he made two important points. First, the necessary calculations are quite simple, involving only distances, angles, and time. Second, unlike desktop computers that work by taking a single piece of information through sequential steps, the insect's brain processes many pieces of information simultaneously, through a number of parallel steps. (Parallel computers are available today for tackling complex computational problems.) This greatly speeds up the data processing, allowing more manipulations to be done with less equipment. This explanation appeals to my scientific mind, but I still marvel at this remarkable example of superminiaturization in our living world.

Having looked at some of the smallest brains among living animals I would like to finish the chapter by considering the brains of dinosaurs. But how do paleontologists study dinosaur brains when nothing is left of the original gray matter? In theory the task is not so formidable as it sounds. Vertebrates' brains are enclosed inside the cranium, and in birds and mammals the fit is so tight that the inside walls conform closely to the shape of the brain. The brain therefore sits snugly inside the cranium like a walnut inside its shell.

A faithful replica of the original brain can therefore be obtained from an avian or mammalian skull using rubber latex. The procedure is simple. First, the various holes in the cranium for nerves and blood vessels are plugged with plasticine. Then, while holding the skull vertically, liquid latex is poured in through the foramen magnum—the large opening at the back of the skull through which the spinal cord passed to the brain. The skull is given a few swirls to make sure all the inside walls of the cranium are coated, then it is left for a few hours for the latex to set. The latex is then peeled away from the inside of the skull, using forceps, and pulled out through the foramen magnum. The finished product, called an endocast, is a faithful replica of the original brain.

The reptilian brain has a far looser fit inside the cranium than a mammal's or a bird's; consequently a modern lizard's endocast is much larger than the actual brain. Another complication is that much of a reptile's cranium is formed of cartilage rather than bone. Since cartilage is seldom preserved in fossils, the uncertainty of our knowledge of the brains of dinosaurs and their Mesozoic relatives increases. However, by making due allowances for these shortcomings, we have been able to make some reasonable estimates of brain sizes and shapes for a number of dinosaurs.

Compared with modern reptiles, many dinosaurs had average-size brains relative to their size. *Tyrannosaurus*, for example, had a brain that was as big relative to its estimated 7-ton body weight as that of a modern reptile. In terms of Jerison's encephalization quotient, then, *Tyrannosaurus* has an EQ close to 1. Similar values have been obtained for some hadrosaurs, but the stegosaurs and horned dinosaurs were far less well endowed, with EQs of only 0.2 for both *Stegosaurus* and *Triceratops*. Sauropods had brains that were half as small again, *Diplodocus* and *Brachiosaurus* having EQs of only 0.1. At the other end of the spectrum are small theropod dinosaurs like *Troodon* (formerly called *Stenonychosaurus*) with an EQ of 5, which is midway in relative size between that of modern reptiles and birds. This small (6.5 feet; 2 meters) dinosaur was lightly built, had sharp claws on its hands and feet, and had enormous eyes that were directed forward, giving it a well-developed

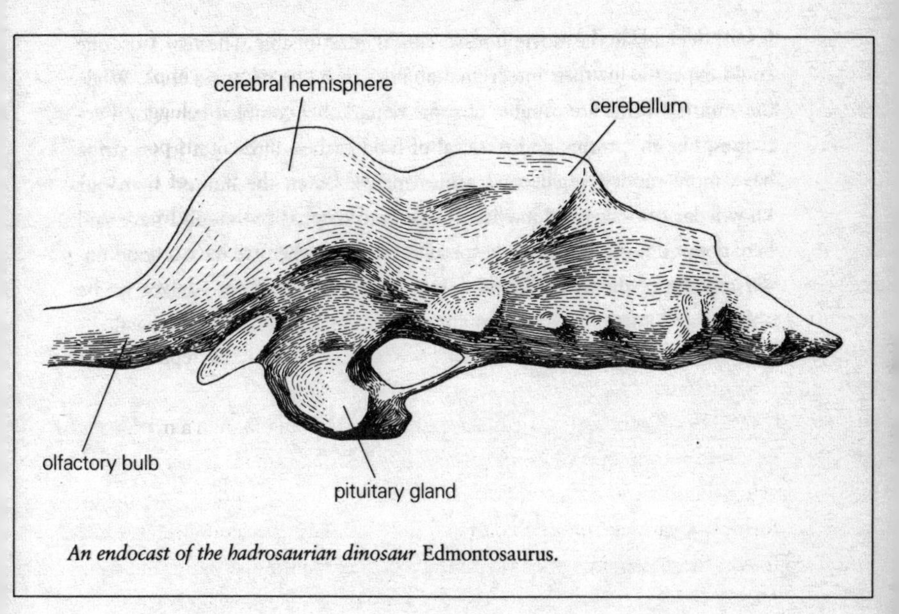

An endocast of the hadrosaurian dinosaur Edmontosaurus.

binocular vision. All of these features suggest that *Troodon* was an active predator.

Although modern reptiles are not noted for their intellectual prowess, many of them are capable of some fairly complex behavioral activities. Crocodiles, for example, participate in elaborate courtship displays, cooperative hunting and feeding activities, and care of their young. All of this is achieved with a typically reptilian-sized brain. Similar levels of complexity could be expected for dinosaurs like *Tyrannosaurus* and its carnivorous relatives, and for the hadrosaurs. The stegosaurs and ceratopsians may have led less complex lives, while the sauropods may have functioned as little more than automatons. Small, seemingly fleet-footed predators (like *Troodon*) and their relatives the dromaeosaurs (which includes *Velociraptor* and *Deinonychus*) may have been more birdlike in their behavioral repertoires than the other dinosaurs. But whether they were capable of the high level of intelligent behavior depicted in the movie *Jurassic Park* is questionable. The very most one could expect is for their intellectual abilities to be on an avian level. While the smartest birds are capable of some remarkably complex behavioral activities, like the storing and retrieval of food caches, birds of a lesser stripe have more modest intellectual achievements. Given the limitation in our knowledge of brains and intellect, I hasten to emphasize that all I have said here about dinosaurs is entirely speculative. Only when we have a good understanding of the living brain—a long way into the future—might we be able to say something more substantive about brains of the remote past.

DRAG IN THE MATERIAL WORLD

Admiral Lord Nelson's battle plan for his engagement with the combined French and Spanish fleets on the morning of October 21, 1805, was as daring as it was novel. Two lines of ships would close with the enemy off Cape Trafalgar, sailing at right angles to the superior naval force. Nelson, aboard the *Victory*, would lead one line, while Vice-Admiral Collingwood, his flag flying from the *Royal Sovereign*, would lead the other. The English ships, unable to bring their guns to bear until they rounded on the enemy, would have to face the concentrated fire of cannon throughout their approach. Speed was of the essence. The English had the advantage of the weather gage and each ship carried her full press of sail. The *Royal Sovereign*, her bottom recently re-coppered, reached the enemy first, but in spite of the advantage of a clean hull, the top speed of a three-decked ship of the line was ponderously slow. Little wonder. They were broad in the beam, bluntly pointed in the bow, and flat at the stern. Contrast this with the sleek lines of the clipper ships that plied the trade routes to the Americas. Built for speed, the secret of the clippers' success was the low drag generated by their hulls.

Drag is the term used for the resistance that a body experiences when moving through a fluid, whether air or water. We are usually unaware of drag in our everyday lives and pay it little heed, but it can be a remarkably large force, requiring expenditures of much energy, and therefore expense, to overcome. You do not notice any drag when walking because you move too slowly. When you run you feel the air against your face, but you are still traveling too slowly to be affected by the drag. Riding a bicycle increases your speed to the point where you feel the drag force, especially when speeding down a hill. If you crouch low over the handlebars you can decrease the drag by reducing your effective size, called the *frontal area*, thereby increasing your speed. Motorcyclists like my editor are well aware of the magnitude of drag forces, and if he were foolish enough to stand up while traveling at speed, he would likely be swept off his machine.

An easy way to experience drag is to extend an arm from a speeding car. Trailing a lazy arm from a slow boat would be equally effective because of the greater density of water over air. Drag, then, increases with speed, fluid density, and object size. Understanding drag requires insight into how fluids behave. Since living organisms experience the same fluid forces as inanimate objects, let us look first at the relatively simple world of the nonliving.

Imagine a small boat gliding slowly across a mirror-smooth lake. The silky water flows smoothly and gently down its sides, each layer slipping past the next without any swirling movements. Such flow is called *laminar*, and the drag forces exerted on the boat are minimal. The outboard motor, which has been barely ticking over, is now throttled open and the boat leaps forward, water swirling and boiling down both sides. This disorganized flow of water is described as *turbulent*, and the drag force on the boat has increased dramatically. Most bodies generate swirling movements when they move through a fluid. Because the energy that sets the fluid particles in motion comes from the body itself, this swirling creates most of the resistance to forward motion. If a body could move through a fluid without causing such disturbances, the drag forces would be greatly diminished.

A useful way of examining drag forces is to visualize what happens when a thin plate is placed in a moving fluid. (Most laboratory experiments on fluid flow have the object stationary and the fluid in motion, as in experiments conducted in wind-tunnels. This arrangement is purely a matter of convenience, and it makes no difference to the end results. All that matters is that the flow of fluid is *relative* to the object being tested.) The most turbulence

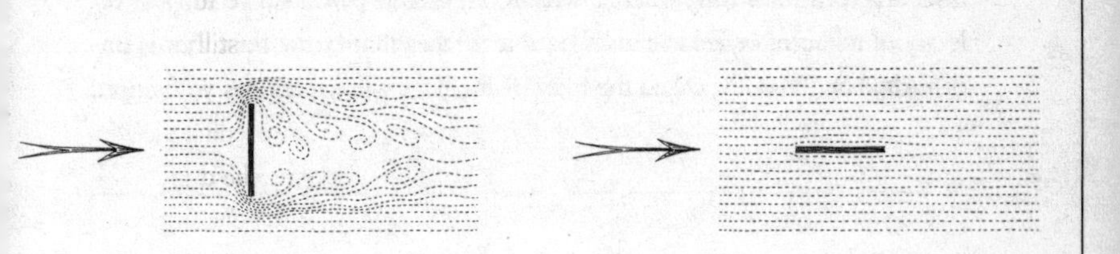

*When a thin plate is placed perpendicular to the direction of flow of a fluid (*left*), it generates considerable turbulence, and drag forces are maximal. When placed parallel to the direction of flow (*right*) the drag forces acting on the plate are minimal.*

is achieved when the plate is placed perpendicular to the flow's direction. Here the frontal area is the area of the entire plate. The fluid can be thought of as being made up of numerous small particles, some of which strike the plate's leading surface and thus oppose forward motion head-on. Many more are diverted before they can reach the plate; some move inward, and create swirling eddies called *vortices*. Extensive vortices can be seen downstream of the plate. Energy is required to move each fluid particle around the plate, and also to accelerate them in new directions. How much energy this requires depends on the particle's inertia—the tendency for it to stay where it is—which is proportionate to its mass and therefore to the fluid's density (mass per unit volume). This component of drag is referred to as *pressure drag* or *form drag*. Almost all the resistance experienced by the perpendicular plate is due to pressure drag.

A second drag component, called *friction drag*, or *skin friction*, acts on the surface of the body, parallel to the flow's direction. Friction drag is caused by a property of fluids called *viscosity*. Viscosity may be thought of as a measure of how readily one layer of fluid slips past an adjacent layer. In general, the lower the viscosity, the easier the flow. Pancake syrup, for example, is far more viscous than water, which is why it takes longer to pour syrup out of a jug than water.

When a fluid flows over a body, the layer in immediate contact with the body attaches to the surface and is therefore stationary. Above this stationary layer is a transition zone where adjacent layers slide past each other. The velocity of adjacent layers increases until it reaches that of the fluid that is undisturbed by the body, called the free stream. This whole region of transition,

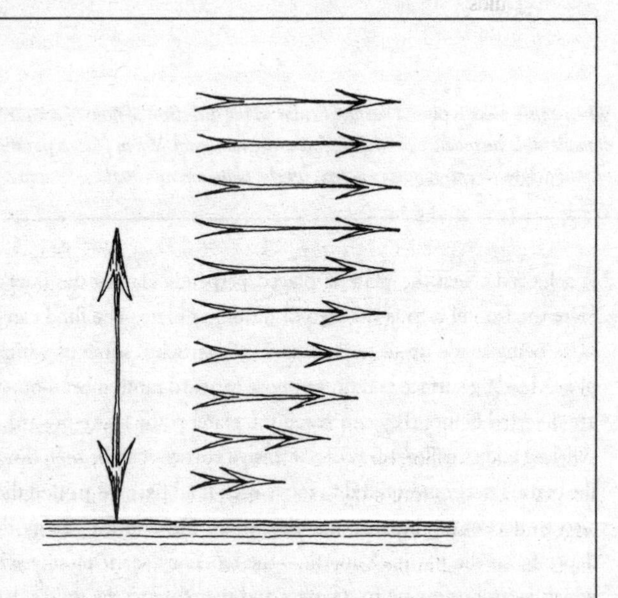

When a fluid flows over a body the layer in contact with the surface is stationary. Above this, the adjacent layers move at increasingly higher velocities until the full velocity of the free fluid stream is reached. The zone of increasing velocities is called the boundary layer.

from the surface to the free-flowing fluid, is called the *boundary layer*. You have probably encountered a boundary layer without realizing it. Do you recall lying flat on a beach, unaware of a cool breeze blowing until you sat up? This is because you were lying in the boundary layer, where the wind velocity was at its lowest.

Adjacent layers of fluid in the boundary layer slip past one another, and each layer moves relatively faster than the one nearest the body. The ease with which adjacent layers slip past one another—or *shear*—varies with the fluid's viscosity. If an imaginary fluid had no viscosity at all, it would have no boundary layer because there would be no resistance to shearing between the layers of fluid, which would all be traveling at the full speed of the free fluid stream. Fluids with low viscosities have thinner boundary layers, so if two objects of the same length were moving at the same speed in two different fluids, the one moving in the lowest-viscosity fluid would have the thinnest boundary layer. Needless to say, if there were no viscosity there would be no friction drag.

To return to the experiment with the thin plate: When it is placed perpendicular to the fluid flow, friction drag is negligible. This is because barely any surface area is parallel to the free stream so almost all resistance to forward motion is due to pressure drag. The reverse holds true when the plate is placed parallel to the flow, minimizing the frontal area. The fluid flow is now almost entirely laminar, and most of the resistance to forward motion is attributable to friction drag.

The total drag acting on a body is a combination of pressure drag and friction drag, attributable to the fluid's density and viscosity, respectively. The sizes of these two forces depend on the body's size and the flow's speed. A large and a small body therefore experience different forces even though they might be moving in the same fluid at the same speed. The relation between these variables is expressed by a ratio called the *Reynolds number* (abbreviated Re) after Osborne Reynolds, a British engineer and physicist.

$$\text{Re} = \frac{\text{length} \times \text{speed} \times \text{density}}{\text{viscosity}}$$

where density and viscosity refer to the fluid, and speed and length refer to the body moving through it. While the length term is usually taken to be body length for swimmers, for fliers it is more usual to use the width of the wing (called the chord). When calculating Reynolds numbers all of the variables have to be measured in the same units, and, because they cancel out in the equation, Reynolds numbers have no units. Although I have been using imperial, or English, units in this book (and metric units parenthetically), metric units are almost universally used in science, so I will use them to determine Reynolds numbers. In metric units, the density of water is 1000 kilograms/cubic meter and the viscosity of water (at 10° C) is 1.304×10^{-3} newton-second per meter squared ($N.s/m^2$). Suppose that we wanted to calculate the Reynolds number of a 2-meter (6.5-foot) long dolphin moving along at 5 meters per second (11 miles per hour). The Reynolds number of the dolphin at this speed is:

$$\frac{2 \times 5 \times 1000}{1.304 \times 10^{-3}} = 7.7 \times 10^6$$

which is quite high. How about a smaller animal, say a 25-millimeter (1-inch) guppy swimming in a fish tank? We already know that small animals move more slowly than large animals, and a reasonable speed for a guppy would be about 20 millimeters (0.75 inches) per second.

$$Re = \frac{0.025 \times 0.020 \times 1000}{1.304 \times 10^{-3}} = 383$$

which is quite low. Very small bodies, like dust particles and small planktonic organisms, move at Reynolds numbers of less than one. Movements at such low Reynolds numbers are dominated by viscous forces, and the greatest drag component is therefore friction drag. This is irrespective of whether the fluid has a high or low viscosity. Air, for example, has a low viscosity, but dust particles illuminated by a shaft of light in a darkened room appear to be floating in syrup. At high Reynolds numbers the situation is reversed: pressure drag is the main force, and viscosity (hence friction drag) is of secondary importance. But friction drag is always present and can never be ignored.

I teach a marine biology course most summers and I do a simple experiment that demonstrates how large the frictional component of drag can be. When we are out on the boat, cruising along at about 4 knots (4.6 miles per hour), we let out about 100 yards (92 meters) of thin line that trails through the water at a high Reynolds number ($Re = 1.5 \times 10^8$). The frontal area of the line is small so pressure drag is negligible. But the surface area parallel to the flow is relatively large, and most of the drag is due to friction. I ask the students to think viscosity when they grab hold of the line, and they are always amazed at how hard they have to pull to overcome the drag force.

In addition to the forces changing with Reynolds numbers, so does the thickness of the boundary layer. This correspondence is evident every time you drive a car in the rain. At low speeds the Reynolds number is low and the boundary layer is correspondingly thick. The beads of water that form on the windows lie within the boundary layer and therefore are not exposed to the full force of the air's free flow; they are not displaced by the air flowing over the car. As the car accelerates, the boundary layer shrinks until it is thinner than the beads of water. At this point the beads that project beyond the boundary layer are exposed to the full force of the free airflow and are swept back by it. The largest droplets are the first to move because they project farthest. You have to drive very fast before the smallest ones are swept away.

There is no clear cut-off point between high and low Reynolds numbers, and many organisms (and vehicles) operate over a range of values. Reynolds numbers in the thousands are high, those in the hundreds and tens are fairly low, and those less than one are definitely low. The following approximations will give you some idea of how they vary among different fliers and swimmers: Jet aircraft and nuclear submarines, 10^8 (one hundred million); large whales, 10^7 (ten million); sharks, 10^6 (one million); albatrosses, 16,000; hummingbirds, 8,000; water beetles, 5,000; dragonflies, 2,000; guppies, 300; protozoa, less than 1; and bacteria, 10^{-4} (one ten-thousandth) and less.

Movements at low and high Reynolds numbers are so fundamentally different that organisms living within the two extremes face distinct suites of problems. As a consequence they usually possess different structural adaptations. The streamlined shape, for example, reduces pressure drag. Since this is a minor component of drag at low Reynolds numbers, the streamlined shape is primarily confined to organisms (and man-made devices) that operate at high Reynolds numbers. By and large, low Reynolds numbers are associated with small body size, so in discussing swimming and flying at low Reynolds numbers I will focus on the minutiae of life in chapter 9. The reverse is true for swimming and flying at high Reynolds numbers, which I discuss in chapter 8, where you will encounter some veritable giants. Those chapters will focus on the organisms themselves, so I will confine the remainder of this chapter to the physics of their adaptations. Most of what follows pertains to the high Reynolds number situation (Re higher than 1,000), starting with the streamlined body.

If a circular plate is placed in a fluid, perpendicular to direction of flow, the vortices generated—hence the pressure drag—is maximal. Replacing the disc with a sphere of the same diameter improves the situation, approximately halving the drag. The flow is essentially laminar around the front or *leading surface* of the sphere and continues like this beyond the widest part of the sphere, called the *shoulder*, toward the back or *trailing surface*. Behind the sphere the fluid flow becomes turbulent, and numerous vortices are generated.

If the leading surface of the sphere is drawn out into a rounded point and the trailing surface is gently tapered, the vortices can be essentially eliminated

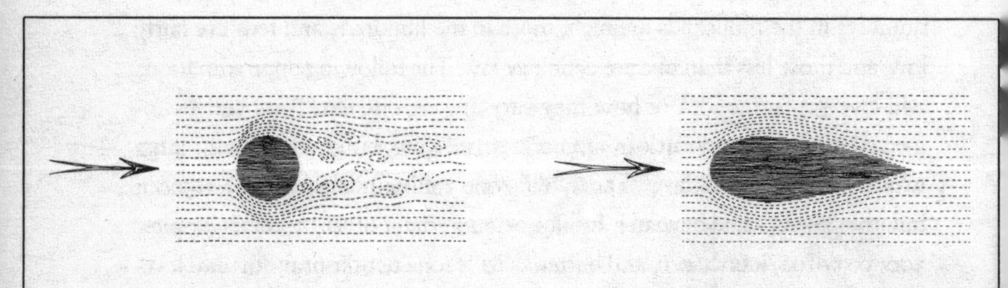

When a sphere (left) *is placed in a flow of fluid it generates less turbulence than a flat plate of the same diameter and therefore experiences smaller drag forces. A streamlined body form* (right) *generates minimal turbulence and therefore minimal drag forces.*

and the resistance is reduced to about 5 percent of that of the perpendicular disc. This torpedo shape is called a *streamlined body*. The optimum streamlined shape for minimizing total drag has the shoulder placed between about one-third and one-half of the way back from the front of the body.

A useful way of expressing the relative slenderness of a streamlined body is by the ratio of the length of the body divided by its diameter. This is called the *fineness ratio*, or slenderness ratio. Increasing the fineness ratio lowers the pressure drag but also increases the surface area and hence the friction drag. An optimum fineness ratio in which the total drag is minimal corresponds to a value of about 4.5.

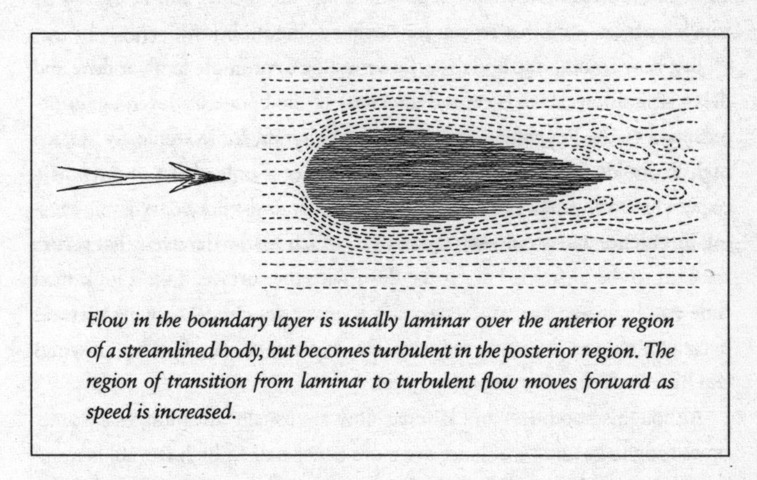

Flow in the boundary layer is usually laminar over the anterior region of a streamlined body, but becomes turbulent in the posterior region. The region of transition from laminar to turbulent flow moves forward as speed is increased.

The major component of drag on a streamlined body is friction, which involves the boundary layer. The fluid flow in the boundary layer, like that in the free-flowing fluid beyond it, can be either laminar or turbulent. If laminar flow could be maintained in the boundary layer, friction drag would be reduced to about 10 percent of what it would be if the flow were turbulent.

The flow in the boundary layer usually starts off being laminar over the first part of the leading surface, but eventually a transition region is reached, beyond which the flow becomes turbulent. The position of the transition region moves forward with greater speed, so as the body moves faster more of the boundary layer becomes turbulent, thereby creating more friction drag. Since surface roughness is one of the factors that causes the boundary layer to become turbulent, having a smooth body is one effective way of reducing friction drag. Competitive sailors take great pains to make the hulls of their sailboats as smooth as possible.

Not only does the boundary layer become turbulent with higher speed, but it also begins to separate from the body's surface. Separation begins at trailing surfaces and moves forward as the body gains speed. The separation in the fluid generates large vortices that reduce, then eliminate, the beneficial effect of the streamline shape. Separation, like turbulence, can be caused by rough surfaces—another reason for having a smooth body.

Just how smooth the body's surface must be to promote laminar flow and delay separation depends on the thickness of the boundary layer, again determined by the Reynolds number. Modern aircraft, for example, fly at such high Reynolds numbers that the boundary layer is only about as thick as a paper-clip. Even small projections extend well into the boundary layer, causing turbulence and promoting separation, which is why the rivets that secure the skin to the airframe have to be flush with the surface. Take a look next time you board an aircraft. Separation is especially critical to wing surfaces because it reduces the lifting capacity. This explains why ice buildup on wings can have such disastrous consequences.

Although smoothness and laminar flow are usually advantageous, sometimes roughness and turbulence are more beneficial. At high Reynolds numbers, for example, a turbulent fluid can remain in contact with a surface for longer than it would if its flow were laminar. This is why golf balls have dimples. The turbulence the dimples generate in the boundary layer delays separation so that the air stays in contact with the surface longer, actually reaching the trailing surface before breaking away. Consequently the width of the wake, hence the drag, is much smaller. A swing that would drive a regular golf ball for 230 yards (212 meters) would drive a smooth one for only about 50 yards (46 meters). The same principle is at work in certain parts of an aircraft, but instead of using dimples, the airflow is set into turbulence by small rectangular projections called vortex generators. They can, for example, be seen on the wings of the Boeing 757.

What about surfaces that are between the two extremes of being parallel or perpendicular to the flow? What forces are experienced by objects moving through a fluid at an angle? When a plate is inclined at a small angle to a moving fluid, such that the leading edge is higher than the trailing edge, it experiences an upthrust or lift force. This is because the plate deflects fluid

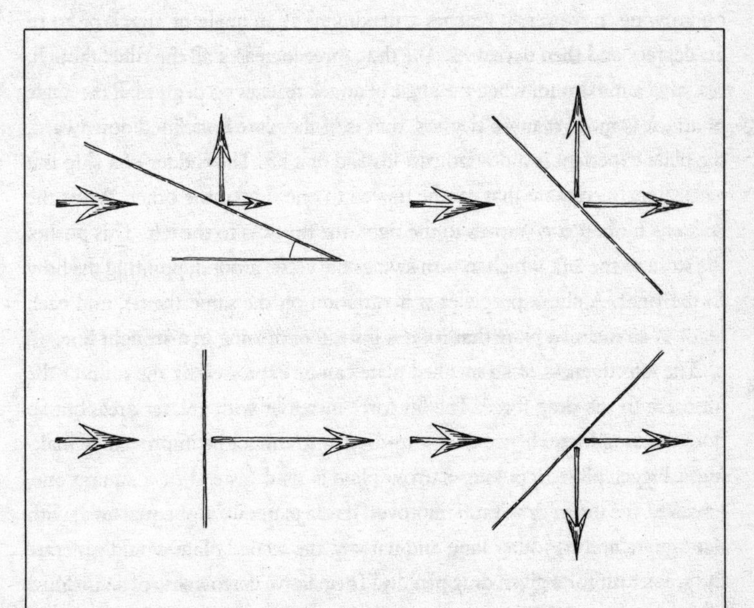

When an inclined plate moves through a fluid at a small angle of attack (top left)*, it experiences upthrust and drag forces. With an increasing angle of attack* (top right) *the magnitude of these forces increases. At right angles* (bottom left) *only drag forces are experienced, while at angles greater than 90 degrees* (bottom right) *the drag forces are accompanied by a downthrust. The magnitude of the forces are indicated by the lengths of the arrows.*

away from its lower surface and the force applied by the fluid has an equal and opposite reaction on the plate. You can demonstrate this by making your hand into an inclined plate and holding it out of the window of a moving car. If you tilt your hand from the horizontal by raising the leading edge, you will feel it being pushed upward by the air.

The angle an inclined plate makes to the direction of flow is called the *angle of attack*. If the angle of attack is gradually increased, the magnitude of the lift force rises—but so does the drag force. The lift force does not keep on growing, however; it reaches a maximum at an angle of attack of 10 to 20 degrees and then decreases. The drag force increases all the time, though, reaching a maximum when the angle of attack reaches 90 degrees. If the angle of attack is more than 90 degrees, that is, if the plate is inclined downward, the plate experiences a downthrust instead of a lift. The rudder of a ship is a vertical inclined plate that can be moved to one side or the other. When the rudder's trailing edge moves to the right, the thrust is to the left. This pushes the stern to the left, which in turn swings the vessel around, pointing the bow to the right. A ship's propeller is a variation on the same theme, and each blade is an inclined plate that rotates instead of moving in a straight line.

The effectiveness of an inclined plate can be expressed by the ratio of the lift force to the drag force. The lift force increases with greater area, but so does the drag force; hence the lift-to-drag ratio cannot be improved by making a bigger plate. If a long narrow plate is used instead of a square one, however, the lift-to-drag ratio improves. If two plates were of equal area, with one square and the other long and narrow, the second plate would generate the greater lift for a given drag penalty. The relative narrowness of an inclined plate is expressed by the *aspect ratio*, the ratio of length to width. A plate 10 units long and 10 units wide has an aspect ratio of 1, whereas one that is 20 units long and 5 wide has an aspect ratio of 4. The lift-to-drag ratio of a plate is also improved by a streamlined profile because, at high Reynolds numbers, this reduces drag.

Most flying machines are heavier than air and require a lift force to keep them airborne. This lift force is usually provided by an airfoil, which takes the form of a wing with a particular kind of curved profile. Although one

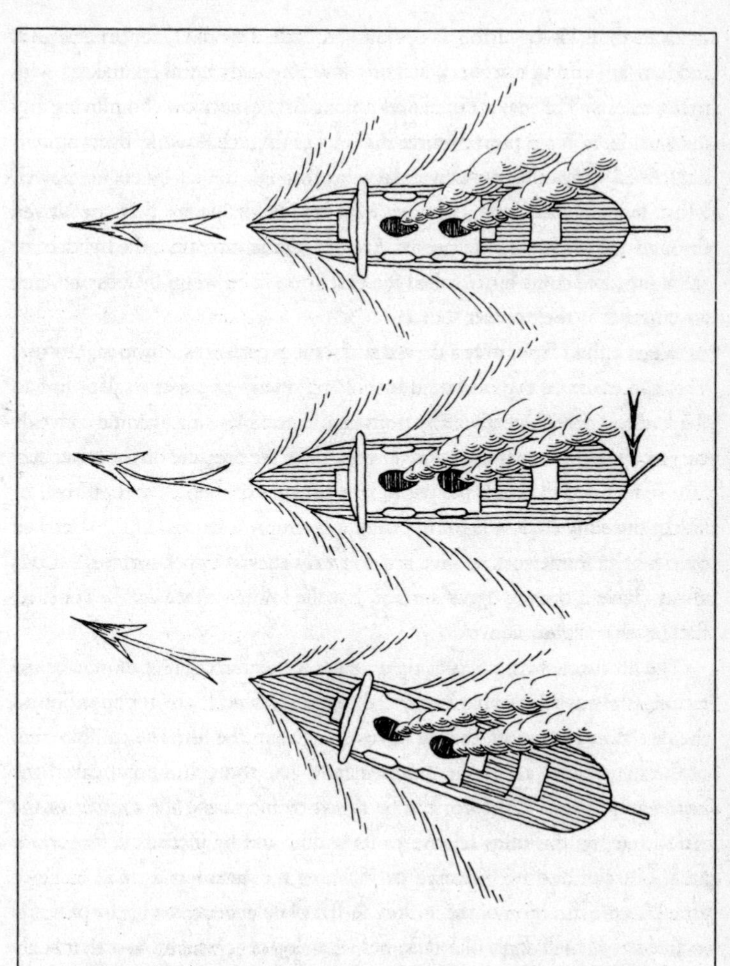

The rudder of a ship functions as an inclined plate. The force generated causes the ship to rotate about its center of mass, pointing the bow in the desired direction. Sailing a straight course (top)*. The rudder is moved so as to turn the ship to the right; the thrust generated by the inclined plate displaces the stern to the left, causing the bow to swing to the right* (middle)*. Sailing on the new course* (bottom)*.*

tends to think of the airfoil as operating at high Reynolds numbers, as with modern aircraft, it can work at fairly low Reynolds numbers too, as with flying insects. The major differences among flying machines, both living and man-made, is in the power source that makes the airfoil work. Most aircraft have fixed wings that are driven forward through the air by engine power. Most flying animals, in contrast, have moveable wings that are driven through the air by muscle power. Even so, some aircraft, called gliders or sailplanes, and some birds, called soarers, fly on fixed wings by using moving air currents as their power source.

When a fluid flows over a curved surface it experiences a drop in pressure. This can easily be demonstrated by holding a strip of paper to your lips so the free end droops gently away from you. If you blow air over the convexly curved upper surface the pressure lowers. Since the pressure on the lower surface remains unchanged, the paper strip airfoil experiences an upthrust, or lift. In the early days of flight the wing was simply a strip of fabric stretched over a light framework to give it a convexly curved upper surface. Airfoils always have a convex upper surface, but the lower surface can be concave, flat, or even slightly convex.

The lift force, which acts at right angles to the direction of motion of the airfoil, is always accompanied by a drag force, which acts in opposition to the direction of motion, that is, at right angles to the lift. The combination of the two forces, called the *total reaction*, acts through a point called the *center of pressure*. Lift force can be raised by increasing the *camber* of the airfoil (i.e., its thickness relative to its width) and by increasing its surface area. Lift can also be increased by inclining the airfoil at a small angle of attack to the direction of the airflow. A flat plate generates a lift force if it is inclined at a small angle of attack, but an airfoil generates lift at a zero angle of attack because of its curved upper surface. Greater lift is always accompanied by greater drag, which requires more energy to push the airfoil through the air. A compromise strategy used in aircraft is to have a variety of high-lift devices to increase the camber and surface area of the wing temporarily, prior to take-off and landing. These take the form of leading and trailing edge flaps that are retracted when the aircraft is in level flight.

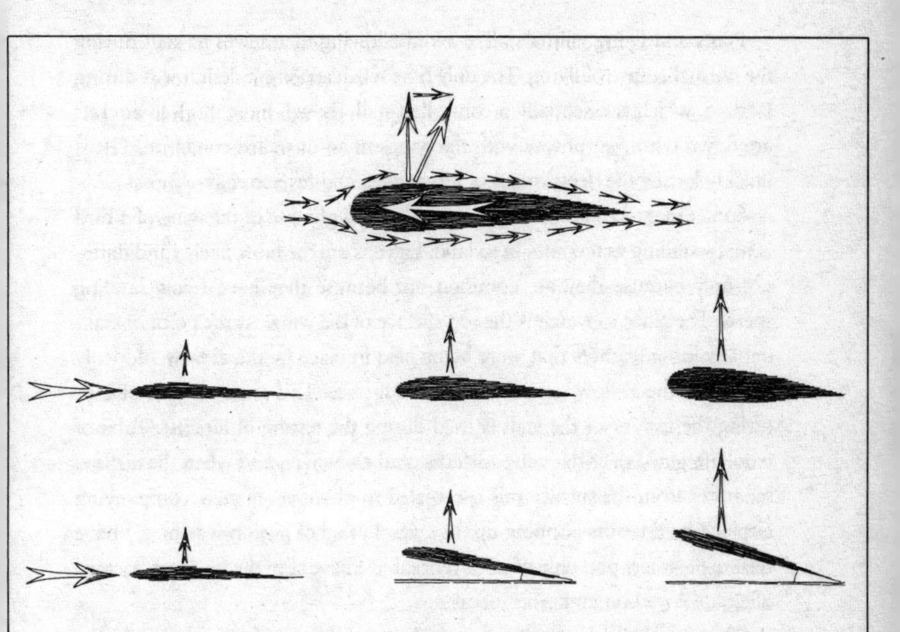

When an airfoil moves through the air it experiences lift and drag forces, which act at right angles to one another. The resultant, the total reaction, acts through a point called the center of pressure (top)*. The lift generated by an airfoil can be increased both by increasing the camber of the airfoil* (middle)*, and by inclining it at a small angle of attack* (bottom)*. Increases in lift, however, are accompanied by increases in drag. The magnitudes of the forces are indicated by the lengths of the arrows.*

As the angle of attack increases the lift rises but soon reaches a maximum value and then falls off rapidly. At this point the airfoil is said to have *stalled*. Stalling can also be brought about by lowering the speed of the air passing over the airfoil—called *airspeed*. The airspeed at which stalling occurs is referred to as the *stalling speed*. Stalling happens when the airflow separates from the airfoil's surface. Separation begins at the trailing edge and moves forward; as it does so, the lift decreases. Most of the lift is lost at the stalling point, and when the lift is no longer large enough to support the weight of the aircraft, it loses height.

Pilots and flying animals alike avoid allowing themselves to stall during the normal course of flying. The only time when stalling is desirable is during landing, which is essentially a controlled stall, though most modern aircraft are flown onto the runway with the wings in an unstalled condition. Lift is quickly lost by the deployment of air-spoilers and reverse engine thrust.

Sometimes, if you are lucky, you can catch a glimpse of the wing of a bird actually stalling as it comes in to land. Pigeons are the most likely candidates, not only because they are common but because they have a low landing speed. The place to watch is the top surface of the wing. At the point of stalling any loose feathers that were being held in place by the airflow suddenly pop up as the airflow separates from the surface. This same method of capturing the instant of the stall is used during the testing of aircraft. Tufts of wool are glued over the wing surfaces, and elsewhere, and when the airflow separates from the surface this is revealed to observers in an accompanying airplane by the tufts popping up. In several years of pigeon-watching I have seen the feathers pop only once or twice, but I have seen the *alula*, or bastard wing, deployed on numerous occasions.

The alula is the bird's thumb, which has the form of a short winglet on the wing's leading edge, about halfway along its length. It is thought that the alula functions as a leading edge slot to delay the onset of separation, and hence to delay stalling. Similar devices are used on the leading edge of aircraft wings. They are called Handley Page leading-edge slots after the aircraft manufacturer who developed them. In its simplest form, as with aircraft built before World War II, the leading edge slot was just a fixed rod attached to the leading edge of the wing, and separated from it by a distance of an inch or so. By re-directing the flow of air over the wing's top surface, the leading-edge slot delays the onset of separation. In later aircraft, including the Messerschmitt 109, the slot was flush-fitting but popped out automatically just before the wing stalled. A bird's alula is similarly deployed just before the wing stalls, which is why it can be seen when a pigeon comes in to land. The alula is small compared with the wing's length, but it is located in the region that is most likely to stall and therefore serves a useful role in spite of its small size.

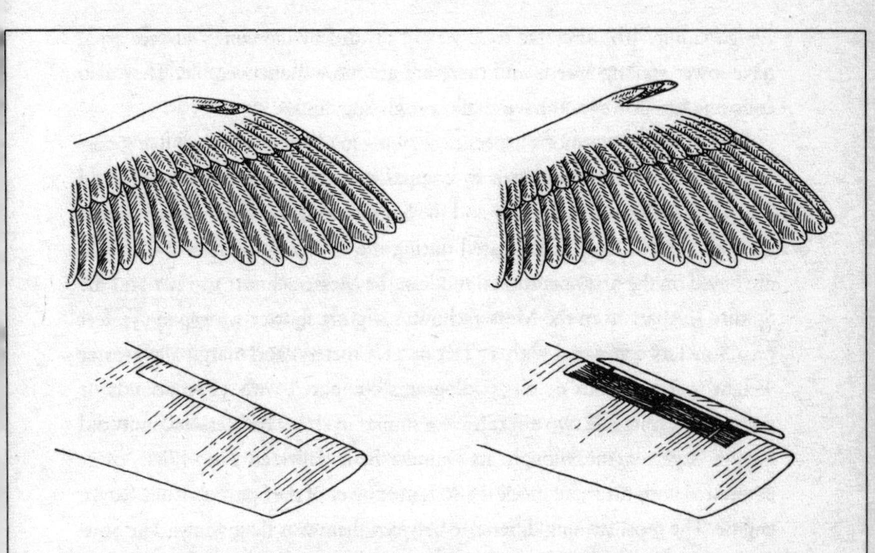

The alula, the bird's thumb, is kept flush with the rest of the wing during normal flying (top left). *Immediately before landing the alula is deployed as a leading edge slot to delay stalling* (top right). *Aircraft have similar anti-stalling devices on the leading edge of the wing* (bottom).

As I mentioned, a useful measure of the efficiency of a wing is the lift-to-drag ratio. Wings with high aspect ratios have a higher lift-to-drag ratio than short wide wings. One of the consequences of this is that the gliding angle (the minimum angle at which the bird can fly) is shallower. Gull wings have higher aspect ratios than pigeon wings, which explains why they can glide at much shallower angles. Having a low gliding angle means that longer distances can be traveled for a given loss of height. Therefore, if a gull and a pigeon glided down to street level from the same height, the gull would travel farther. Wings with high aspect ratios not only have high lift-to-drag ratios but also high stalling speeds, which means that higher speeds must be maintained to prevent stalling.

Weight also has a big influence on flying performance. Aircraft with low *wing loadings* (the aircraft's total weight divided by the wing's surface area) have lower stalling speeds and therefore are more maneuverable. They also consume less power and have shallower gliding angles.

The connections among aspect ratio, wing loading, power, and flying performance can be demonstrated by comparing two famous fighters of World War II: the Messerschmitt 109 and the Supermarine Spitfire. Many different versions of each aircraft appeared during the war, and the data I am using are based on the first operational models: the Messerschmitt 109 E-1 and the Spitfire I. Apart from the Messerschmitt's slightly shorter wingspan (32 feet or 9.8 meters compared with 37 feet or 11.3 meters) and marginally greater weight (5400 pounds or 2455 kilograms compared with 5280 pounds or 2400 kilograms), the two aircraft were similar in size. The Messerschmitt did have a larger engine, though: its Daimler-Benz delivered 1100 horsepower compared with the more modest 880 horsepower of the Spitfire's Rolls Royce engine. The most striking difference between them was their wings. The Spitfire had much broader, elliptical wings, giving a lower aspect ratio and a wing area almost 40 percent greater than that of the Messerschmitt. As a consequence, the wing loading of the Messerschmitt was 33 percent higher than the Spitfire's: 32 pounds per square foot or 154 kilograms per square meter compared with 24 pounds per square foot or 115 kilograms per square meter. This conferred a higher stalling speed on the Messerschmitt. It could climb and dive faster than the Spitfire and had a higher operational ceiling. But because of its higher stalling speed, it could not fly as slowly and thus could not make such tight turns, making it less maneuverable. One of the defense strategies Spitfire pilots used was to put their machines into tight spirals, making it almost impossible for their faster and wider-turning adversaries to get a good shot.

To fly or swim successfully a body must be able to compensate for alterations in its path caused by fluid movements, and it must be able to change course at will. Three basic movements can occur: yaw, pitch, and roll. I will illustrate each by reference to a simple aircraft. Yaw is a side-to-side turning movement about a vertical axis, pitch is a nose up or down movement about a horizontal axis, and roll is a rotation about a longitudinal axis. All three

The Spitfire (top) *had relatively broader and slightly larger wings than the Messerschmitt (*bottom*). Both aircraft weighed about the same. Consequently the Spitfire had a lower wing loading, hence greater maneuverability.*

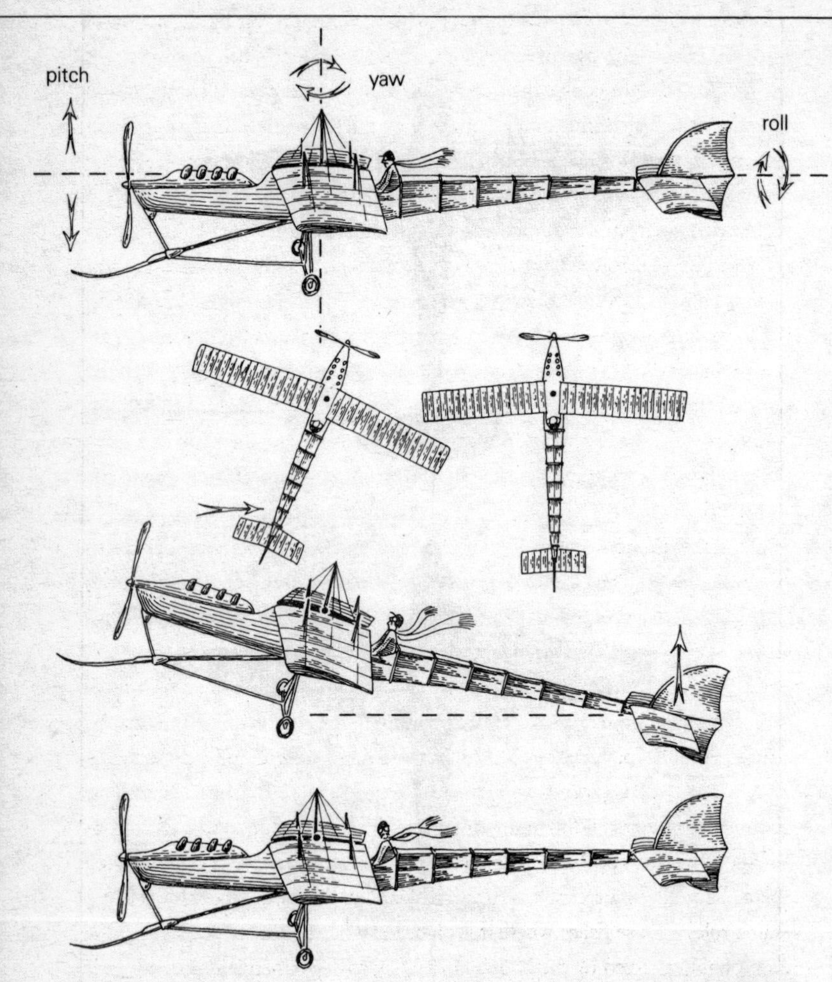

Top: *Movements in the three planes of space are pitch, yaw, and roll.* Second row: *When an aircraft yaws* (left) *the rudder becomes an inclined plate and the thrust generated swings the tail back in line again.* Third row: *When the nose pitches up, the horizontal stabilizer becomes an inclined plate and the lift generated raises the tail, returning the aircraft to level flight* (bottom).

movements are corrected for during the flight of an aircraft by the action of inclined plates. Yaw is corrected by the vertical rudder. When an airplane is flying on a straight course the rudder cuts through the air in the direction of travel, but when the machine begins to yaw the rudder becomes an inclined plate set at a small angle of attack and the deflection force brings the airplane back on a straight course. Control of pitch is brought about by the action of the horizontal stabilizer, which acts like an inclined plate. The flight feathers of a dart or an arrow correct for yaw and pitch in the same manner; if a dart is thrown with a wobble it corrects itself rapidly and flies straight. In addition to correcting for yaw and pitch an airplane's tail has adjustable control surfaces that can *cause* it to yaw (turn) and pitch (climb or dive). Roll is effected by the ailerons, which are elongated flaps on the trailing edges of the wings, placed toward the wing tips. Increasing the angle of attack of an aileron gives an additional lift to the wing on that side, making the aircraft roll toward the other side.

To illustrate how swimmers handle yaw, pitch, and roll, consider a fish such as a herring. But I must begin with the caveat that my explanation is an idealized and oversimplified account of the true situation. A swimming fish is far more complex than a flying airplane. Unlike the man-made structure, each part of the fish's body is in motion; consequently the control surfaces are not static, although I treat them as though they were. Like most fishes, the herring has paired fore- and hindfins, called pectoral and pelvic fins. A single dorsal fin is placed at the thickest part of the body (some fishes have more than one dorsal fin) and a single anal fin is ventral to, and just in front of, the tail. Yaw is probably corrected for by the anal fin's acting as an inclined plate, as with an airplane's vertical rudder. The dorsal fin probably has the same role in those fishes where it is placed behind the center of gravity. Yaw is probably effected by the body's lateral flexing movements and also by extending one of the pectoral fins more than the other; the increased drag on the more fully extended fin swings the head toward that side. The pelvic fins probably correct for pitch, much like the horizontal stabilizers of an airplane. When it comes to effecting pitch, though, the pectoral fins act like the hydroplanes of a submarine. Inclining the pectoral fins upward generates a lift

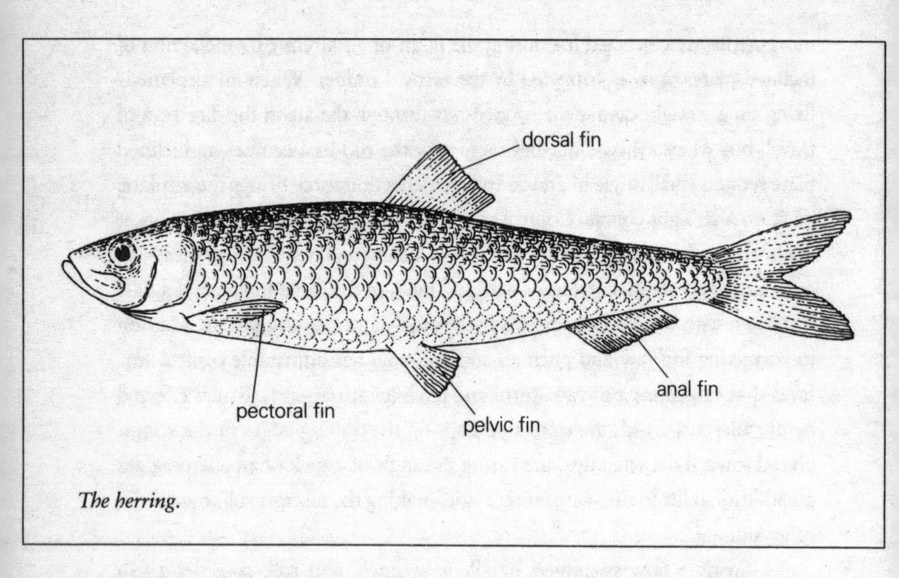

The herring.

force that raises the head, while the reverse is achieved by tilting them downward. Roll could be induced by generating different lift forces in the right and left fins, either by partially extending or retracting one of the fins, or by setting them at different angles of attack. Roll is probably corrected for by the damping effect of the dorsal fin "biting" into the water as the fish rolls to one side.

Stability is the automatic correction that occurs when a body deviates from a straight course. Control, in contrast, is the action a body takes to change course. The two variables are inversely related, so an object that is especially stable is more difficult to turn. An airplane with a particularly large tail, or a fish with a large dorsal fin set well behind its center of gravity, would need larger forces to change their directions than less stable bodies. Some fishes have dorsal fins set well forward, partly or wholly in front of the center of gravity. Placed in this forward position the fin is destabilizing. When a fish turns, say toward the right, the fin acts as an inclined plate in front of the center of gravity, generating a force that pushes the head even farther toward the right. The forward-placed dorsal fin thereafter enables the fish to turn more easily, just as a person sitting at the front of a canoe can steer more easily than the person sitting at the back.

To summarize: Movements at high Reynolds numbers are dominated by inertial forces and these forces are utilized to produce forward thrust. Swimmers achieve this using a variety of devices including tails, fins, and body movements, whereas fliers use only wings. The major drag component is pressure drag, attributed to inertial forces, but friction drag, attributable to viscous forces, still has to be taken into account even at the highest Reynolds numbers. The relative magnitudes of the inertial and viscous forces, and the size and speed of the object, are expressed by the Reynolds number. Objects moving at similar Reynolds numbers experience similar forces and can be considered to be equivalent, regardless of the fluid in which they move. Thus a helium balloon rising through air and a marble falling through syrup would experience similar forces and behave in similar ways if their respective Reynolds numbers were similar. The streamline, a torpedo-shaped device for reducing drag, is only functional at high Reynolds numbers. It can be seen in a variety of applications from the cross-sectional shapes of fishes' fins and birds' bodies to the shape of energy-efficient cars.

HIGH FLIERS

DARWIN WAS A meticulous observer, both on land and at sea, and his five-year voyage around the world aboard the *Beagle* afforded him a unique opportunity to chronicle nature's richness. One November morning in 1832, when the *Beagle* was in the mouth of the Plata River, he noticed that the rigging was frosted with webs and swarming with thousands of tiny spiders. The weather was fine and clear, with a gentle offshore breeze. Although the land was sixty miles distant, the air was filled with terrestrial flotsam of single strands of silk, each carrying a parachuting spider.

Several months later Darwin was making observations on the flight of the condor.

The condors may oftentimes be seen at a great height, soaring over a certain spot in the most graceful circles. Except when rising from the ground, I do not recollect ever having seen one of these birds flap its wings. . . . I watched several for nearly half an hour, without once taking off my eyes; they moved in large curves, sweeping in circles, descending and ascending without a single flap.

These two observations graphically illustrate the two ends of a continuous spectrum between low and high Reynolds number fliers. The minute spider, carried on the breeze by a gossamer thread, capitalizes on drag forces in this high viscosity situation, while the condor uses lift forces in a world dominated by inertia.

The condor, one of the largest living fliers, provides a perfect example of flight at high Reynolds numbers. Condors have a wingspan of almost 10 feet (3 meters), a weight of 25 pounds (11.3 kilograms), and have difficulty taking off, especially after eating large quantities of meat. They have to take a run at it, flapping their wings vigorously all the while. Once airborne they are magnificent fliers. Darwin noted that hunters exploited the condor's takeoff difficulties to capture them. The hunters would lay out a carcass on a flat piece of ground and build a low fence of sticks around it. They would then wait for the condors to come and gorge themselves, after which the birds would be unable to escape because they lacked enough space to make a takeoff run. The unwitting condors were seemingly victims of their large size rather than their gluttony. But, while large birds generally do have greater difficulties than small ones in getting airborne, there is no simple causal relation between body mass and takeoff performance. The North American wild turkey (*Meleagris gallopavo*), for example, is as heavy as a condor but can take off almost vertically from a standing start.

Before considering the reasons for size-related differences in flying performance, I want to explain how wings function. All heavier-than-air flying machines must generate two forces: lift, to overcome gravity; and forward thrust, to overcome drag. Powered airplanes with fixed wings use engines to overcome drag and wings to provide lift, but animals' wings perform both functions, using muscle power to drive them. While differing in detail among species, a bird's flapping flight mechanism involves a downward and forward power stroke with the wing fully extended, followed by an upward and back-

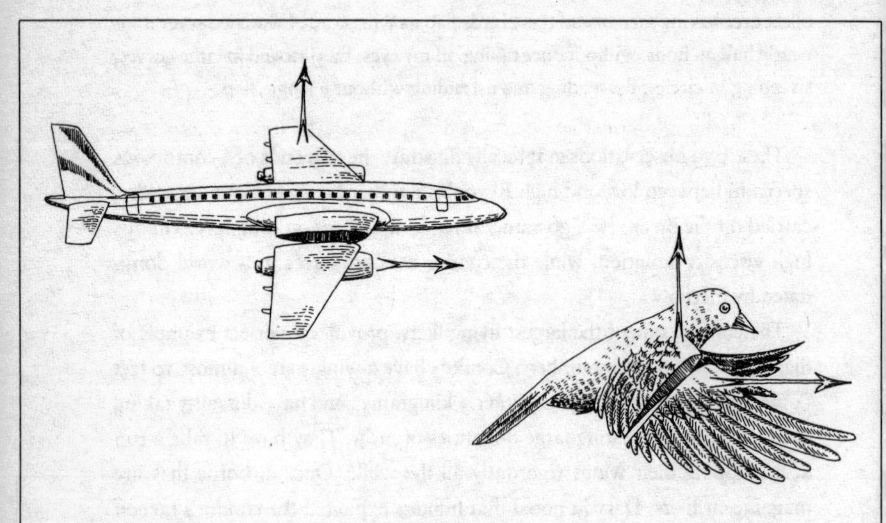

Powered aircraft use engines to overcome drag and propel them forward, and wings to generate lift. A bird's wing has to perform both functions.

ward recovery stroke with the wing partially retracted. The wing has an airfoil cross-section, and as this is driven through the air during the power stroke, it generates a lift force directed obliquely forward and upward, counteracting both drag and gravity. The upward wing beat is largely a recovery stroke, though most birds probably generate some lift and thrust when flying slowly, such as when they take off. The muscle power for these movements is mostly provided by the breast muscles attached to the sternum, a deeply keeled structure in flying birds.

Imagine carving the breast meat from a turkey. The first few slices carve easily—nice clean slices that look well on the platter. But cut another slice and an elongate hole appears, right in the middle of an otherwise perfect slice, because there are two breast muscles on each side of the sternum, one lying immediately beneath the other. The outer one is called the pectoralis and is much larger than the underlying supracoracoideus muscle. The pectoralis muscle powers the downstroke, while the supracoracoideus, usually only about one-tenth as large, powers the recovery stroke. These two muscles can effect diametrically opposed wing movements while occupying similar positions because of how they attach to the humerus. The insertion tendon of the pectoralis muscle connects directly to the leading edge of the upper part of the humerus, so that when the muscle contracts it pulls the wing downward and forward. The tendon of the supracoracoideus, in contrast, loops through a bony canal in the shoulder region, and is inserted on the top of the humerus. Consequently, when the smaller muscle contracts, it elevates the wing in preparation for the next downstroke.

Once a bird is flying its forward motion is added to the forward motion of its wings during the power stroke; this increases the airspeed over the wing.

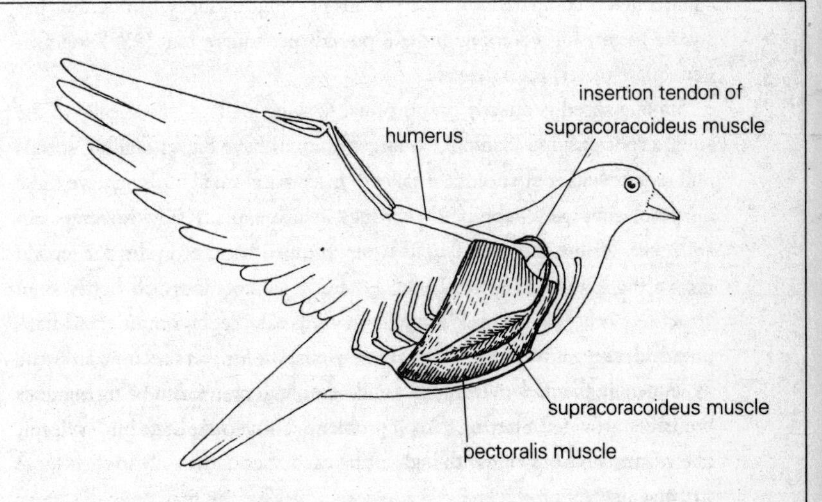

The two major muscles powering the bird's wing are the pectoralis, which produces the downstroke, and the supracoracoideus, which produces the upstroke. The supracoracoideus lies beneath the pectoralis, and its tendon of insertion passes through a bony canal to attach to the top surface of the humerus.

Prior to takeoff, though, the airspeed over its wings is zero (unless the bird is heading into the wind), and lift can be achieved only by raising this airspeed above the stalling speed. Small fliers, like small birds, bats, and insects, have no difficulty with this, as evidenced by how readily houseflies and sparrows take to the air. Most large fliers, though, have to run or jump while flapping their wings to become airborne. These differences in flight performance are directly attributable to the differences in size.

As things get larger their volumes and masses increase with the cube of their length, whereas their areas increase only with the square. Consequently large animals have higher wing loadings than do small ones. (Remember that wing loading equals mass divided by the wing's surface area, or $length^3$/$length^2$, therefore wing loading is proportional to length.) For example, the wing loading of a bee is about 3.3 ounces per square foot (1 kilogram per square meter) compared with 1.1 pounds per square foot (5.1 kilograms per square meter) for a starling and 1.8 pounds per square foot (8.6 kilograms per square meter) for a pelican.

Stalling speed is directly proportional to wing loading (specifically to the square root of wing loading), so larger animals have higher stalling speeds and require higher airspeeds for takeoff. Insects and small birds achieve these airspeeds simply by flapping their wings, which explains how hoverers can fly. A bee, for instance, can flap its wings at more than 200 cycles per second (hence the low-pitched buzz), but a starling cannot do much better than about five beats per second. Many large birds take off by running and flapping and taking advantage of prevailing winds, while others become airborne by launching themselves from trees and other high places. But being big does not necessarily mean having takeoff problems. Some fairly large birds of prey, like vultures, have wings with high lifting capacities due mainly to their large size and shape.

Conventional wisdom has it that the power requirement for sustained flight rises at a rate slightly higher than the increase in body mass, whereas the muscle power available to a bird changes according to body $mass^{2/3}$. It follows that sustained flight has an upper size limit, and this corresponds to a body mass of about 26 pounds (12 kilograms). Several living birds are of this size, including the Andean condor (*Vultur gryphus*), mute swan (*Cygnus olor*), white pelican (*Pelecanus onocrotalus*), wild turkey (*Meleagris gallopavo*), and the kori bustard (*Ardeotis kori*).

Like so many other good ideas proposed by biologists to explain nature, this one was bound to be ruined by uncaring animals. In this instance the group comprised 147 insects, 10 birds, and 7 bats. James Marden, a zoologist at Pennsylvania State University, carefully designed an experiment in which each of these animals was progressively loaded with attached weights to determine the maximum load each could lift. Calculating the maximum force generated by an individual's flight muscles was a simple matter of multiplying the maximum all-up weight (weight of animal plus the maximum weight it could lift off the ground) by the acceleration due to gravity (which the animal had to overcome). When a double logarithmic graph of maximum lift force was plotted against flight muscle mass, the result was a straight-line graph with a gradient of 1. This meant that the force generated per unit weight of flight muscle was constant and independent of body mass. So a bird weighing 200 grams would be able to generate twice as much lifting force as one weighing 100 grams provided the ratio of flight muscle mass to body mass remained the same. Although these experimental results measured lift force rather than power output, Marden argued that this serves as an index to power output. But there is one caveat—the largest animal in the study was a pigeon, though a later study did include a hawk. It is possible that the relation might not hold for large animals, but the fact that it applies to such a wide diversity of smaller ones suggests that it might. Although there is insufficient evidence to make any definitive statements, the conventional wisdom that flight performance is limited by the scaling relationship between required and available muscle power looks to be on shaky ground.

Marden's loading experiment showed a remarkable degree of constancy between muscle force and muscle mass over a wide range of animals. The individual species, however, varied in their takeoff ability, primarily owing to differences in the ratio of flight muscle mass to body mass. Animals with large amounts of flight muscle are much more adept at taking off than those at the other end of the scale. They simply have larger engines. The coot, a member of the rail family noted for its poor flying ability, has flight muscles that comprise only about 13 percent of its body mass, compared with the pigeon's 22 percent. The coot, which lives beside lakes and ponds, has to take a long flapping run across the water to become airborne, but the pigeon can take off vertically from a standing start. Condors similarly have a small mass of flight muscles relative to their large body mass. But they only need enough muscle power to become airborne; then they can exploit thermals to stay aloft. It would be an unnecessary metabolic expense, not to mention an excess of weight, to have more flight muscle than was needed for takeoff. Data for a variety of bats, birds, and insects show that the minimum mass of flight muscle for standing-start takeoffs is about 16 percent of the body mass.

Once airborne, sustained flapping flight requires that an animal's flight muscles respire aerobically, because an oxygen debt can be accumulated only for short periods. The maximum power output for aerobic muscles is about 100 watts per kilogram, and this is true for vertebrates and insects. As we saw earlier, anaerobic muscles (white) generate more power than aerobic ones (red), and for birds the factor is about 2 to 2.5. The factor is even higher for reptiles—up to about 4.5—and this has implications for the giant pterosaurs that I will deal with later on. Because of these differences in power output, fliers can maximize their takeoff performance by having relatively higher percentages of white fibers (fast glycolytic cells) in their flight muscles. This is the strategy of the wild turkey and its game bird relatives. Endurance fliers like ducks, in contrast, have higher percentages of red muscle fibers (slow oxidative cells). Insects apparently do not have any option because their flight muscles have little or no anaerobic capacity.

To recap: As animals get larger their wing loadings increase and their wing-beat frequencies decrease. Flying performance is not entirely dictated by size and there is a strong correlation with the percentage of flying muscle to body mass. An obvious size trend does exist, however. At one end of the spectrum are small maneuverable flappers, like hummingbirds, that can hover but not soar. Middle-sized flappers like pigeons take off readily and can soar but not hover. Large birds, like pelicans, soar with only intermittent flapping. They are not very maneuverable and often have difficulty taking off.

While power output from flight muscles may not impose such a low size ceiling as previously supposed, there may be structural limitations on how large fliers can become. As animals become larger the stresses on their bones increase with the cube of their length, whereas the ability to resist these forces increases only with the square. Terrestrial animals have the option of strengthening their limb bones by narrowing their marrow cavities, as with heavy animals like elephants, which have solid bones. Flying animals, though, do not have this option because they must restrict their weight. Aeronautical engineers were faced with the same problem, which they resolved by inventing new materials and new technologies. The Wright brothers' first biplane, made of wood, wire, and canvas, was driven by a piston engine and weighed 605 pounds (275 kilograms). The largest aircraft flying today, the Antonov An-225, is constructed of lightweight alloys, plastics, and composites, is driven by gas turbines, and has a staggering weight of 591 tons. Technology has emancipated the airplane from the constraints of size, but living organisms are still made of the same materials and powered by the same muscles that evolved millions of years ago. These constraints ensure that paleontologists will never dig up the remains of a flying animal as large as a DC9, far less a 747.

I teach a course in vertebrate mechanics, and the section I enjoy most is the one on flight. We have fun experimenting with models in wind tunnels—streamlines, airfoils, and the like—and then we look at bird wings. I can get very enthusiastic about the wings of modern aircraft with their leading edge slots and multi-slatted trailing edges, but these are crude devices compared with the aerodynamic elegance of a bird's wing. Just to hold a bird's wing—to feel its lightness and stiffness, check its camber, and see how the feathers glide past each other without leaving gaps when the wing extends and retracts—fills me with wonder. We could not build a variable geometry wing as light and stiff and aerodynamically sound, so how is this achieved in a bird? The secret lies in the structure of the feather.

Try holding a feather in your fingers, but find a regular stiff, contour feather, not one of those fluffy downy feathers used to stuff pillows. If you hold the feather by the quill, which is the lower end of the rachis, or shaft, you will notice that it is round and hollow. If you try squeezing it, it will not

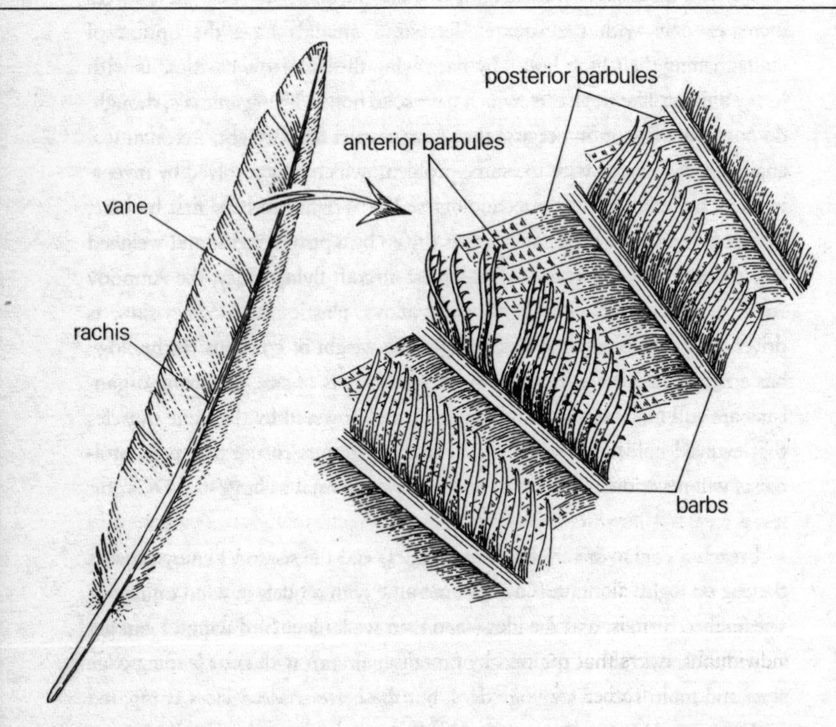

The structure of a bird's feather combines lightweight construction with stiffness. The body of the feather, the vane, comprises a series of barbs that branch off from the central rachis. The barbs, in turn, have side branches called barbules. The anterior barbules bear hooks and the posterior ones have ridged plates onto which the hooks latch, like Velcro.

collapse. Tubes are strong. While still holding the shaft between your finger and thumb, check its stiffness by applying gentle pressure with another finger. It is remarkably stiff. The material from which the shaft and the rest of the feather is made is keratin—the stuff of hair and nail—a tough, light protein. Notice that the feather's body, the *vane*, is formed from numerous fore-and-aft side branches from the shaft, called *barbs*. Adjacent barbs are connected together along their entire length. Run a finger along the vane and you will see that it is quite flexible and will distort. The barbs remain locked together, though, so when you remove your finger the vane springs back to its original shape. The barbs can be separated by pulling them apart, which tears the vane. You can reverse this by stroking the vane between two fingers. Birds do this for themselves when preening their feathers with their bills. This works because of the way the barbs are joined together. Examine the barbs with a lens and note their side branches, called *barbules*. The anterior barbules are armed with hooks, while the posterior ones have ridged plates onto which the hooks of the adjacent barbules latch, just like Velcro. The eloquence of the design, which combines stiffness and strength with lightweight construction, can be appreciated by holding a feather up to the light and seeing how much empty space there is. Indeed, the large volume of air entrapped by feathers accounts for their excellent insulating properties.

The vane has an airfoil cross-section that is convex on the upper surface and concave below. Furthermore, the wing's contour feathers have their shafts displaced toward the leading edge, just as the main spar of an aircraft wing is displaced toward the leading edge. This configuration aligns the supporting structure along the center of lift and prevents twisting during flight. The feather, then, is a mini-wing; when a wing is in flight air often flows over individual feathers that respond by generating lift. You can check the airfoil properties of a feather for yourself. Lay a feather on the table right-side up (convex surface up) and attach the quill to the table with a small piece of tape so the feather hinges freely. With the feather flat on the table, gently blow air over the feather parallel with the surface of the table. The lift generated causes the feather to pop up.

The same design principles of minimizing weight while maximizing

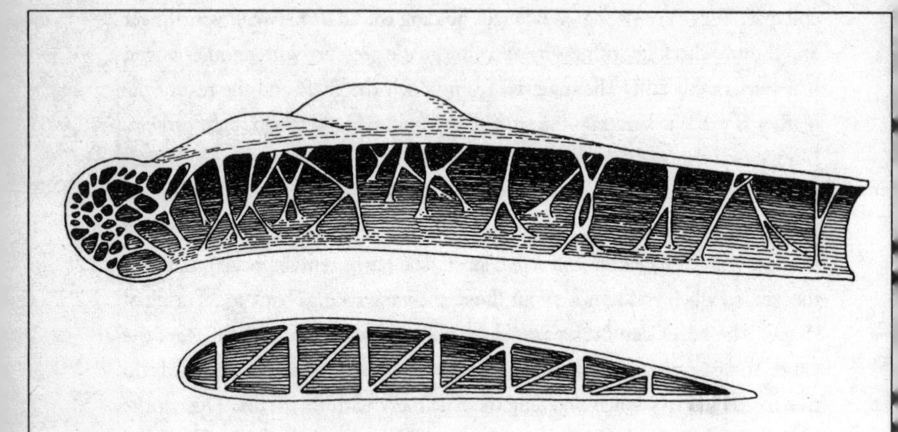

The bones of a bird's wing combine strength and lightness by having thin walls, braced by internal struts (top)*. A similar strategy is used in aircraft construction (*bottom*).*

strength and stiffness obtains for the construction of the wing skeleton. The long bones making up the wing—the humerus, radius, and ulna—are thin-walled tubes with thin internal struts in the regions subjected to the greatest stresses. Experiments on flying bats have shown that torsional (twisting) stresses are considerably higher in the humerus and radius of fliers than of terrestrial animals, because the wing bones are positioned near the leading edge whereas the wing's center of lift lies farther back. This positioning tends to twist the wing during flight. Torsional forces are best resisted by wide tubes—another functional reason for the tubular construction of wing bones. While the bony skeleton forms the wing's most anterior part, much of the leading edge is formed by an elastic web of skin, the propatagium, that extends from the shoulder to the wrist. The next time you prepare poultry for the oven, note how large and remarkably resilient it is. The propatagium probably functions as a leading edge slot and is supplied with muscles that might be able to modify its shape.

Aircraft have devices for measuring airspeed and attitude, giving the pilot

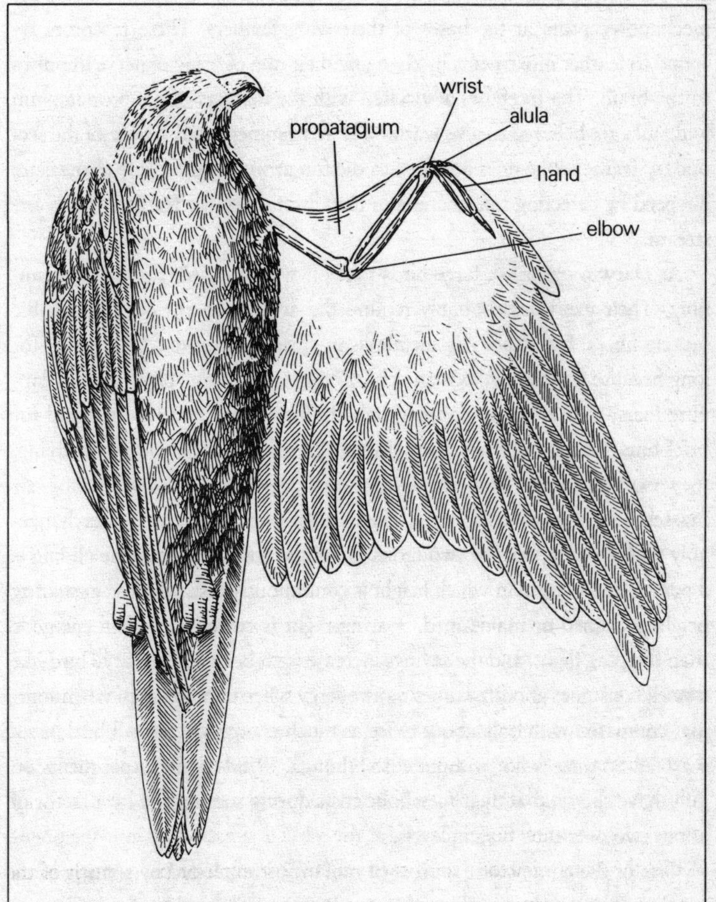

Much of the leading edge of a bird's wing is formed by an elastic web of skin called the propatagium.

warning of the separation of the boundary layer from the wing's upper surface prior to stalling. Birds appear to have mechanical devices in the form of mechanoreceptors at the bases of their wing feathers. These receptors respond to feather movements by changing their rate of firing of nerve impulses to the brain. The receptors associated with the feathers of the propatagium and alula are believed to give warning of an imminent stall. Those of the secondary feathers (the ones attached to the forearm) might be able to measure airspeed by detecting rate changes in the vibration of the feathers in the airstream.

As Darwin observed, large birds take off with a lot of flapping and running. Their exertions probably require the additional power of anaerobic muscle fibers, but anaerobic metabolism (sprinting) cannot be kept up for long because it incurs an oxygen debt. Consequently large birds are not flapping fliers; they resort to flapping only during takeoff and landing, and for brief bursts during normal flying. Instead of using muscle energy for flying, they extract energy from the environment, specifically from moving air masses, a strategy called *soaring*. The term *gliding* is often used interchangeably with soaring, but the two terms are not synonymous because gliding is a passive mechanism in which height is continuously lost, whereas in soaring height is gained or maintained. Soaring flight is considerably less energetic than flapping flight, and the savings increase with body size. A large bird like a stork consumes about 23 times more energy when flapping than when soaring, compared with only about twice as much energy for a small bird like a warbler. Soaring is not without costs, though. Wind-tunnel experiments on gulls have shown that their metabolic costs during soaring rise by a factor of about two over their resting levels.

One of the commonest sources of moving air exploited by soaring birds are thermals, which are rising columns of warm air caused by the differential warming rates of different substrates. A large rock in the middle of a savannah, for example, will warm up beneath the sun more rapidly than the surrounding grassland, sending up a vertical column of warm air. Thermals are usually quite narrow, often only a few feet in diameter, and fliers have to make tight turns to stay within them. Such maneuverability requires slow flying speeds, as does turning tight corners in a car. Soaring birds that exploit thermals therefore require low wing loadings. Most soarers also have low-aspect-ratio wings because these have lower stalling speeds than do high-aspect-ratio wings of the same area.

Soarers also employ a high-lift device that Darwin referred to in his description of the condor: "As they glided close over my head, I intently watched . . . the outlines of the separate and great terminal feathers of each wing; and these separate feathers . . . were seen distinct against the blue sky." These separated feathers, which give the wing tips a ragged appearance, are particularly noticeable in birds of prey and in crows and ravens. Quite often these terminal feathers are *emarginated*—that is, the outer portion is narrower than the inner portion—so the wing tip is permanently digitate. This type of wing tip, referred to as *slotted*, reduces drag during flight, and thereby raises the wing's lift-to-drag ratio. This device also aids takeoff because when the wings are flapped the individual feathers function as individual high-aspect-ratio mini-wings that contribute to lift and improve the lift-to-drag ratio of the wing. This avian high-lift device attracted the attention of aeronautical engineers when the oil crisis hit the western world. They experimented with finlets attached to the wing tips of a Boeing 707. The results were encouraging: Fuel consumption dropped and takeoff performance improved. Finlets are now a standard feature of several modern airliners, including the Boeing 747.

Thermals are not restricted to narrow vertical columns because they can be displaced horizontally by winds. This displacement gives rise to soarable corridors, called *thermal streets*, that birds can use for cross-country flights. Thermals can also take the form of large ascending bubbles, up to about a mile (2 kilometers) in diameter. As long as a flier stays within the bubble it can fly in any direction and still gain height. A strategy used by soaring birds and glider pilots is to ascend on a thermal to gain height, leave it and glide across country, and then ascend on another one. Birds are as adept at recognizing likely places for thermals as experienced glider pilots; this was graphically demonstrated to Colin Pennycuick, one of the foremost authorities on bird flight. While studying soaring birds in East Africa with a glider,

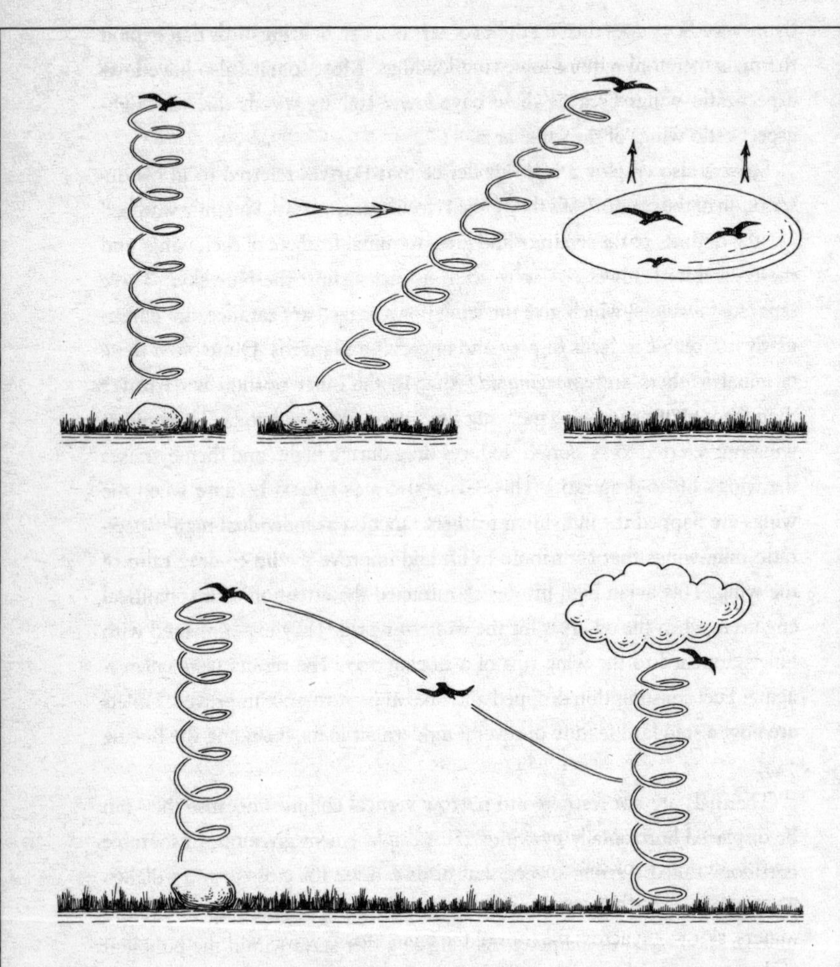

Rising columns of warm air, called thermals, *are formed when the sun heats the ground unevenly, as when a rock warms up faster than the surrounding grassland* (top left). *Thermals can be displaced horizontally by winds to form soarable corridors called* thermal streets (top middle). *Thermals can also take the form of large ascending bubbles* (top right). *Birds and gliders can make cross-country flights by gaining height in one thermal, and then gliding off to another to gain height again* (bottom).

he noticed that their flight paths took them through more thermals than would be expected if they had been flying randomly. On one particular flight he decided to follow a vulture, and made much better time by doing so than on his return trip when he had to rely on his own piloting skills!

An entirely different source of moving air exploited by soarers, primarily birds, arises when horizontal winds are deflected upward, as when onshore winds blow against sea cliffs. As the air masses are not confined, maneuverability is of less importance. The primary consideration is efficiency—maximizing the lift-to-drag ratio. This is achieved by employing high-aspect-ratio wings like those of gulls, which are naturally associated with sea cliffs. Their long slender wings have a higher lift-to-drag ratio than the much lower-aspect-ratio wings of the pigeon. As a consequence, gulls can glide at much shallower angles than pigeons.

The kind of soaring exemplified by gulls and practiced by many other sea birds is described as slope soaring. Slope soaring is not confined to cliffs because when the wind blows against waves, especially when a heavy sea is running, it generates updrafts that seabirds can exploit. A few years ago I was on a Canadian research vessel fishing for swordfish off the coast of Nova Scotia. Gulls were the most common birds at first, but they became scarce when we got farther out to sea. Most of our fishing was done on the edge of the continental shelf, about 100 miles from shore. When the wind began to blow and the waves began to grow along came the petrels and shearwaters and fulmars—members of the *procellariiform* or tube-nosed birds. Storm petrels, the smallest ones, are sparrow-sized and fly close to the waves with a skittish fluttering of their wings, sometimes hovering, sometimes dabbling with their feet on the water. But the larger procellariiforms are adept slope-soarers, skimming across the tops of the wind-blown waves with barely a flap of their gracile wings.

The albatross is the largest of the procellariiforms. The largest species, the wandering albatross, has a wingspan of 11 feet (3.4 meters), which is the largest of all living birds. They commonly slope-soar, like their smaller relatives, but they also practice dynamic soaring, exploiting the differences in the horizontal wind velocities at different heights above the sea—that is, the boundary layer. The albatross makes a shallow dive into the wind from a height of about 60 feet (20 meters). As it loses altitude it gains ground speed (speed relative to the ground). At the bottom of the descent, just above sea level, the bird makes a turn into the wind and its ground speed is so high that it ascends rapidly. As it climbs higher its ground speed slows, but because the air encountered is moving progressively faster, its airspeed does not diminish much. As a consequence the bird continues to climb, even though its ground speed is low, until it gets high enough to do it again. The extensive wanderings of the albatross across the trackless southern oceans are almost legendary. Although it has long been surmised that they make extensive feeding forays during the breeding season, this has now been verified by satellite tracking. Each member of a breeding pair takes turns looking after the nest while the other heads off to sea in search of food. During the investigation, conducted on a breeding colony in the Crozet Islands in the Indian Ocean, radio transmitters were attached to five males and their movements were tracked by satellite. Their feeding forays, lasting two to five weeks, covered distances of 2,257 miles (3,664 kilometers) to a remarkable 9,363 miles (15,200 kilometers). The flying was mostly done in daylight, at speeds up to 49 miles per hour (80 kilometers per hour), but the birds were still active at night, especially during moonlight. Most of the time was spent on the wing, though they did land on the ocean's surface for brief rests that never exceeded an hour and a half.

A few years ago the Royal Ontario Museum obtained an albatross from New Zealand and I showed it to my class. Its body was pretty impressive—about a yard long and as plump as a goose—but when we spread out the wings the tips almost touched the walls of our small laboratory. The students were suitably impressed, as much by the wings' slenderness as by their length. Later, when the internal organs had been removed, I examined the stomach and found it crammed full with horny beaks of squid. How many lonely miles of the South Pacific had that magnificent bird patrolled to collect such a bountiful harvest?

High-aspect-ratio wings are energy efficient; they enable an albatross to cover enormous distances at minimal cost. The disadvantage is that, for a given wing loading, they have higher stalling speeds, which requires higher takeoff and landing speeds. The albatross's situation is exacerbated by its high wing-loading, due to its large size; this detrimental effect on takeoffs and landings may have prompted the epithet "gooney bird." As depicted in wildlife movies, albatrosses approach their takeoffs with the apprehensions of a first-time flyer. Facing into the wind they commence their run, flapping their long unwieldy wings as fast as they can, while trying hard not to trip over their huge webbed feet. Sometimes they have to abort their takeoff at the last moment, tumbling into an undignified heap of wings and feet at the end of the runway. But once airborne they are transformed into a rhapsody of motion, soaring and wheeling on unmoving wings with all the grace of a ballerina. Their high landing speeds rob them of maneuverability, and their return to terra firma can be as undignified as their departure. Condors, like other large birds with low-aspect-ratio wings, have much lower landing speeds and can usually alight upon the ground without incident. Their possession of slotted wing-tips—universally absent from the slender-winged marine soarers—may help to smooth their landings.

Thermal soaring can take place over land only when there are differences in surface temperatures, which happens only during daytime, under the sun's influence. Grounded at night, soaring birds must wait until the sun heats the land before they can take to the air. The first thermals of the day are weak, and can only support birds with the lowest wing loadings—the smallest ones. This was clearly demonstrated by the meticulous observations of E. H. Hankin who wrote a wonderful book on bird flight at the beginning of the century. Hankin, who lived near a meat-processing plant in India, where the smells attracted carrion-feeding birds from far and wide, went to great lengths to record the birds' flying performances, aspect ratios, and wing loadings. He observed that as the early morning sun strengthened and thermals formed, the smallest soarers flapped their wings and took to the skies. These were followed by larger birds until finally the largest ones were able to soar. Landings at day's end were repeated in reverse order; the smallest birds were the last to be grounded by the failing thermal systems.

Before leaving the topic of soaring flight, I should mention the frigate bird because its flying performance seems most closely similar to that of the pterosaurs, an extinct group of reptiles.

Frigate birds are a familiar sight in the tropics, their scimitar wings jet-black against the sky. Though they are large birds, as big as a pelican, they are more lightly built and have much narrower wings. Pennycuick compared the flying performances of frigate birds and pelicans and provided sample wing dimensions. The frigate bird's wingspan is 7.5 feet (2.3 meters), similar to the pelican's 6.8 feet (2.1 meters), but its aspect ratio was 12.8 compared with 9.8. In spite of its much higher aspect ratio, the frigate bird's wing loading was about half that of the pelican (0.77 pounds per square foot [3.7 kilograms per square meter] compared with 1.2 pounds/foot^2 [5.9 kilograms/meter^2] for the pelican). This combination of a low wing loading and a high aspect ratio is probably unique among birds. It confers remarkable maneuverability with a shallow gliding angle.

I have spent many hours watching frigate birds in the Galápagos Islands and was once treated to a display of the air piracy that has earned them the name "man-o'-war birds." Owing to their superior flying performance they literally fly rings around other birds. They harass their unfortunate victims so mercilessly that the victims disgorge the fish they have caught and the frigate birds deftly swoop and catch the fish in mid-air.

Soaring land birds are grounded at night, but not marine birds, certainly not in the tropics anyway. There the trade winds usually have a different temperature from the sea, and the result is convection currents, hence thermals, day and night. Since frigate birds are so maneuverable, they can exploit these thermals and fly night and day. Compared with the pelican, the frigate bird's only disadvantage is its inability to take off from the sea, a consequence of its higher-aspect-ratio wings. It has been said that frigate birds always land in trees because they cannot take off from the ground. The frigate birds I saw certainly did land and take off from trees and shrubs, but the shrubs were sometimes no more than a foot high—hardly a height advantage, except perhaps to get above the sluggish part of the boundary layer. The birds had absolutely no difficulties becoming airborne. They simply faced into the onshore breeze, spread their wings, and launched themselves into the air.

Soaring is a requisite of large size because of the excessive energy requirements of flapping flight. For example, a starling expends about nine times

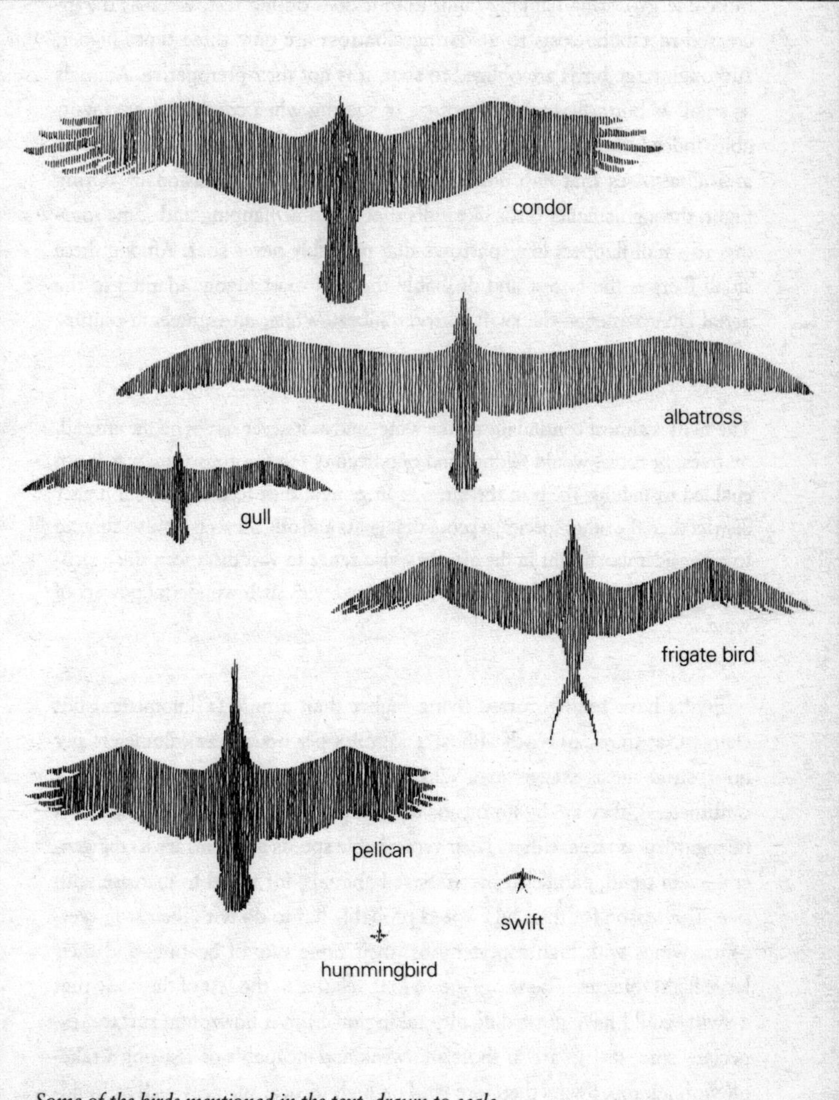

Some of the birds mentioned in the text, drawn to scale.

more energy during flapping flight than it does during rest, whereas the increased metabolic costs to a soaring albatross are only three times higher. Although large birds are obliged to soar, it is not their prerogative. Animals as small as butterflies take advantage of soaring when conditions are favorable. Indeed, there is a continuous spectrum from large soarers like condors and albatrosses that flap during takeoff, landing, and occasionally during flight, through smaller birds like gulls that do some flapping and some soaring, to small flappers like sparrows that probably never soar. Among these small fliers is the fastest and probably the bird most highly adapted to the aerial environment—the swift. Wrote Gilbert White, an eighteenth-century naturalist who lived in the English hamlet of Selborne:

> The swift is almost continually on the wing; and as it never settles on the ground, on trees, or roofs, would seldom find opportunity for amorous rites, was it not enabled to indulge them in the air. . . . In general they feed in a much higher district than the other species; a proof that gnats and other insects do also abound to a considerable height in the air; they also range to vast distances, since locomotion is no labour to them who are endowed with such wonderful powers of wing.

Swifts have been reported flying higher than a mile (2 kilometers) but claims that they can reach almost 200 miles per hour (322 kilometers per hour) strike me as exaggerated. With wingspans ranging up to 9 inches (22.5 centimeters), they are by no means the smallest of birds, but they could not be regarded as large, either. Their remarkable speeds are contrary to the general avian trend, paralleled by terrestrial animals, for speed to increase with size. The reason for their high speed probably has to do with their stiff, crescentic wings with high aspect ratios, their rapid rate of beating, and their large flight muscles. The wings are so long relative to the rest of the body that a swift would have great difficulty taking off from a horizontal surface, especially since its legs are so short and weak and incapable of assisting a takeoff with a jump. Swifts therefore land on high vertical surfaces such as buildings, where they hold on with feet that are well adapted for clinging. They roost with their chest pressed snugly against the vertical surface and take off simply by letting go with their feet and launching themselves into the air.

Closely related to swifts are hummingbirds, noted for their amazing hovering abilities, iridescent colors, and small size. The Cuban bee hummer, the smallest of all birds, is no bigger than a bumble-bee and weighs less than 2 grams; the largest, the giant hummingbird, is little bigger than a sparrow and weighs 20 grams. To put this size range into perspective, many flying insects are larger than hummingbirds, including the goliath beetle—twice as heavy as the biggest hummingbird. Insects excel at hovering, so it is no coincidence

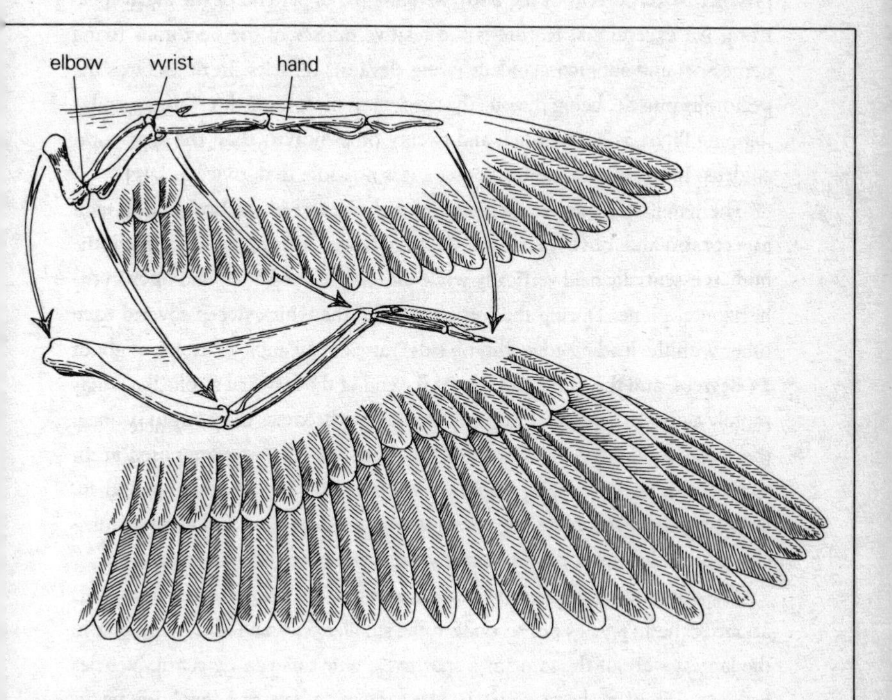

The wing of a hummingbird (top) *differs from that of other birds in being stiff (the only movable joint is at the shoulder) and in being formed mainly by the hand.*

that the flying mechanism of hummingbirds, so radically different from that of all other birds, shares much in common with insect flight.

Hummingbirds differ from other birds in that the wing is formed almost entirely from the hand rather than having a major contribution from the upper arm too. Furthermore, the only movable joint is at the shoulder. The entire wing is stiff and unbending like that of an insect, but it can still be laid flat against the side of the body when the bird is at rest because of the flexibility at the shoulder joint. The sternum, to which the major flight muscles attach, is relatively larger than in other birds. The muscles themselves comprise 25 to 30 percent of the body weight—the upper end of the avian spectrum. An exceptional feature is the relative masses of the pectoralis (wing depressor) and supracoracoideus (wing elevator) muscles. In other birds the pectoralis muscle, being the one that generates most of the lift during regular flapping flight, is between ten and twenty times heavier than the supracoracoideus. In the hummingbird, though, it is no more than twice as large.

The wing, relatively long compared with the rest of the body, has a high aspect ratio and moves in a most remarkable way. During hovering flight the body is essentially held vertically while the wings sweep back and forth in the horizontal plane. During the forward stroke the wings sweep toward each other with the leading edge (thumb side) raised at an angle of attack of about 23 degrees, and this generates lift. At the end of the forward stroke the wings rapidly twist, so what was the top surface now becomes the bottom surface; the wings then sweep backward, again with the leading edge raised at an angle of attack of about 23 degrees to continue lift. Unlike other birds, then, both wing beats generate lift, which is why the depressor and elevator muscles (pectoralis and supracoracoideus) are more nearly the same size.

The wing-beat frequency varies according to the size of the species. It is about 80 hertz (cycles per second) in the smaller ones and about 10 hertz in the largest—about the same as a sparrow's. Our ears can detect frequencies as low as about 50 hertz, so we perceive the wing-beat as a low hum; hence the name hummingbird. The Reynolds numbers for hummingbirds' wings during hovering flight are in the range of 6700 to 9200, compared with 41,000 for a circling frigate bird and 59,000 for a circling pelican.

During hovering flight the hummingbird's body is essentially held vertically and the wings are swept back and forth in the horizontal plane.

Many terrestrial animals, ourselves included, can store some energy generated by a moving limb at the end of its stride, and use it to help power the next stride. Called strain energy, it is mostly stored in the legs' stretched tendons, just as it is stored in the stretched elastic of a sling-shot. For instance, our Achilles tendon stores about 35 percent of the total energy turnover of running during each stride, which is a considerable saving. It had always been thought that flapping birds did not store strain energy, but we now have evidence that hummingbirds may be able to do so.

While feeding, hummingbirds hover over flowers and slip their long bills inside, pumping up nectar with a long tubular tongue. They do not feed entirely on nectar but supplement their sweet tooth with small insects and spiders that they take on the wing. When they want to fly off elsewhere they change their orientation from nearly vertical to horizontal, adjust the sweep of their wings, and zoom off. The wings now beat vertically rather than horizontally, and produce lift on both the up and the down strokes. Forward thrust appears to be limited to the downbeat, though, which explains why the wing depressor (pectoralis muscle) is twice as large as the elevator (supracoracoideus).

Despite their small size, hummingbirds have been clocked at 27 miles per hour (43 kilometers per hour) in wind-tunnel experiments, which is twice as fast as the similarly sized chickadee. They are also remarkable acrobats, and can fly in any direction, including backward. Hovering flight is energetically expensive and is therefore mainly restricted to small fliers like hummingbirds and insects. Helicopters are also fuel expensive and restricted in size; even the biggest ones are small compared with the largest fixed-wing airplanes. The energetic lifestyle of hummingbirds causes them to feed frequently. Even so, they do take time out of their busy schedule to rest, and spend much of their day perching. At night, depending on their nutritional state and on ambient temperatures, they may go into torpor, dropping their body temperature to ambient levels to reduce their metabolic costs.

Most hummingbirds are permanent residents of the tropics. But some species, like the ruby-throated hummingbird, spend their summers in Canada and the United States, migrating a few thousand miles afterward to overwinter in the south. For some, this journey entails traversing the Gulf of Mexico, a sea-crossing of 500 miles (800 kilometers). Without allowing for winds, this journey would take them about a day to complete. Like other migratory birds they prepare for their journey by storing body fat, increasing their body mass by about 60 percent. Most of this fat is metabolized during their journey—probably the most rapid weight loss among vertebrates.

Hummingbirds compete with bats for smallest size. The smallest species, the butterfly bat (*Craseonycteris thonglongyai*), weighs only about 2 grams and has a wingspan of about 5 inches (13 centimeters). The largest bat, the flying fox (*Pteropus giganteus*), has a 6-foot wingspan (almost 2 meters) and weighs about 2.6 pounds (1.2 kilograms). Bats are a fascinating group that has attracted scientific attention for over two centuries. But their public image needs a lot more work. While writing this chapter I attempted to share some of my new-found knowledge of bats with my wife, but she did not want to hear—she does not like bats and that is all there is to it.

Bats are creatures of the night. We can see birds any old day, but our encounters with bats are few and far between. A glimpsed silhouette against the night sky, then nothing. Even if they were abroad during the daytime, I doubt that they would ingratiate themselves with us in the same way that birds do. Birds have bright colors, they sing, and, when they are not flying, they are busy building nests, courting, and preening. Bats, in contrast, usually wear drab colors, sing off-key, and, when not flying, hang upside-down, wrap their membranous wings about them like a cloak, and go to sleep. As if that were not bad enough, most of them are downright ugly! Perhaps it is just as well they are nocturnal because, with so little going for them, they would be horribly persecuted.

Bats can wrap their wings about themselves because their wings are flexible and supported by four jointed fingers. The wing membrane is also attached to their feet—one reason why they walk awkwardly. Most species are incapable of anything more than a clumsy shuffle, and so they avoid landing on the ground. The membrane itself is very thin, often no thicker than a plastic sandwich bag, and is quite tough, fairly resilient, and well supplied with blood vessels. The rich blood supply keeps the living wing membrane well nourished and facilitates the repair of tears. The blood flow through the wings also serves as an effective means of getting rid of excess heat. Muscles are only about 20 percent efficient; consequently, 80 percent of the energy released during their contraction is heat. Flapping flight generates vast quantities of heat that requires disposal. Birds lose some of this heat through their respiratory system, but this is costly with respect to evaporative water loss, so most of their heat is probably lost through the feet. It has been suggested that one of the reasons why birds often fly at high altitudes is to lose heat. The bat's flapping wings are ideal radiators that shed excess heat without evaporative water loss.

The resilient wing membrane is attached to the legs and body, as well as being supported by the fingers, and many bats have muscle strands running through the membrane, too. This arrangement gives bats far more versatility in wing shape than birds enjoy. Like birds they have a propatagium, but its leading edge is muscular. This enables the camber to be changed at will, like the leading-edge slot of an airliner. Unlike birds, bats have a membrane stretched between their legs, the uropatagium. It is extensive in some species and contributes to the wing lift; in others it is reduced to a narrow strip tacked to the inside edges of the legs.

Having so much control over the shape of their wings gives bats consid-

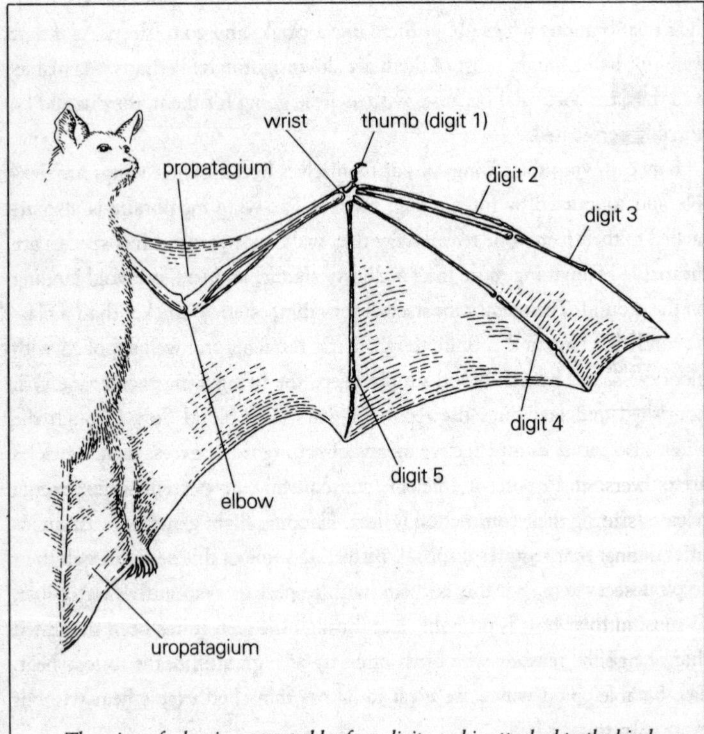

The wing of a bat is supported by four digits and is attached to the hind leg. The propatagium is muscular, and its shape can be changed at will. Most bats have a membrane between their legs, the uropatagium.

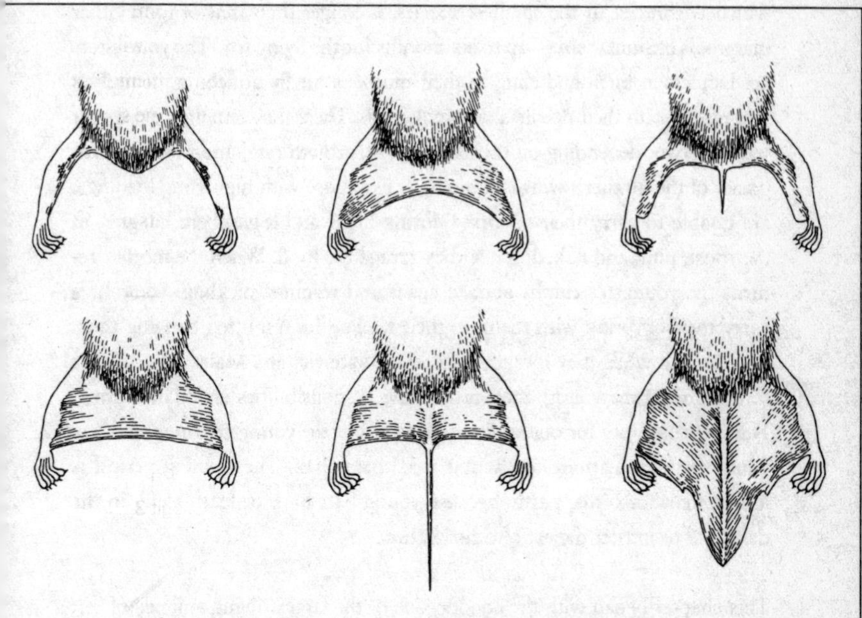

The extent and shape of the wing membrane between the bat's hind legs (the uropatagium) varies considerably.

erable versatility in the air. And having lower wing-loadings than birds of similar size gives them enormous maneuverability. This has been attested to by striking strobe-light photographs of insectivorous species scooping up flying insects with their uropatagium and bending down to pick them up with their mouths. Smaller bats can hover and many of them feed on nectar, like hummingbirds. Larger ones, like the flying foxes that feed on fruit, are probably unable to hover, but they frequently glide for long distances and probably soar, too.

Birds lay eggs and, aside from the time they spend in the oviduct, eggs do not have to be carried during flight. Bats, in contrast, bear their young alive, which adds to the wing loading of pregnant females. The gestation period, which is shortest in the smallest species, is longer than that of most other mammals of similar size—up to six months for the flying fox. The youngsters are helpless at birth and cling to their mother's fur by attaching themselves to a nipple with their needle-sharp milk teeth. There they remain for the next week or two, depending on their initial size, growth rate, and the lifting capacity of the mother's wings. Some bats, like those with high wing-loadings, are unable to carry their offspring during flight and leave them hanging in the roost, pink and naked, while they forage for food. When the mother returns the youngster climbs aboard again and resumes suckling. Some bats carry their offspring with them on their feeding forays, often hanging them on a branch while they forage in the immediate vicinity. Males take no part in the care of the young; their procreative responsibilities ended at mating. Nursing continues for one to three months, but the young continue suckling while they are learning to fly and feed themselves. The training period is longer than for birds, partly because young bats have to learn to fly in the dark and to master the art of echolocation.

This chapter began with the condor, one of the largest flying animals of our modern world. To look at the largest creatures ever to take flight, we need to consider a group of reptiles that became extinct over 65 million years ago—the pterosaurs. Like bats, with which they are frequently and sometimes erroneously compared, pterosaurs had a membranous wing. However, instead of being supported by most of the hand, it was supported only along its leading edge. The thumb, index, and middle fingers are short and end in hooked claws, but the fourth finger is greatly elongated and supported about half the entire wingspan.

Two types of pterosaurs inhabited the world: long-tailed and short-tailed. The long-tailed ones appeared during the Triassic Period, about 220 million years ago, but became extinct toward the end of the Jurassic, at about the time the short-tailed ones appeared. Whether the wing membrane attached to the legs, as in bats, or stopped short at the level of the hips has been a contentious issue among specialists and is still largely unresolved. The different genera probably varied a great deal, as in the bat's uropatagium. The

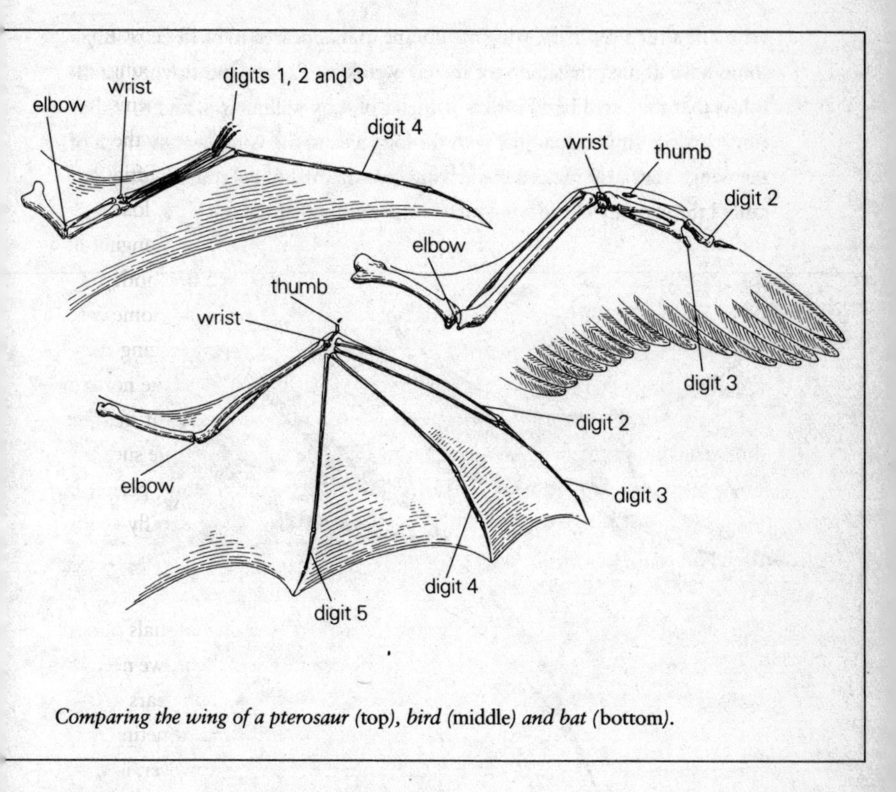

Comparing the wing of a pterosaur (top), *bird* (middle) *and bat* (bottom).

available evidence, which is meager, supports the view that the membrane attached to the feet in long-tailed pterosaurs, but might not have extended any farther than the knee region in some tailless ones. Peter Wellnhoffer follows this general pattern in the illustrations of his magnificent book on pterosaurs. The idea that the wing membrane stopped short of the legs, as championed by Kevin Padian of the University of California at Berkeley, has always presented the problem of how it could have kept from flapping uncontrollably during flight. Nevertheless, Padian, a paleontologist, and Jeremy Rayner of the University of Bristol, a specialist in animal flight, offered a solution. Ever since the latter part of the nineteenth century it has been known that there were fine structures in the wing membrane that appeared to be fibers, though some have argued that they are merely wrinkles. Padian and Rayner established that they are fibers, with a diameter of 0.05 millimeters, and that they run spanwise (that is, parallel with the long axis of the wing) across much of the wing. They also make a convincing case that the fibers could have maintained the stability of the wing membrane during flight.

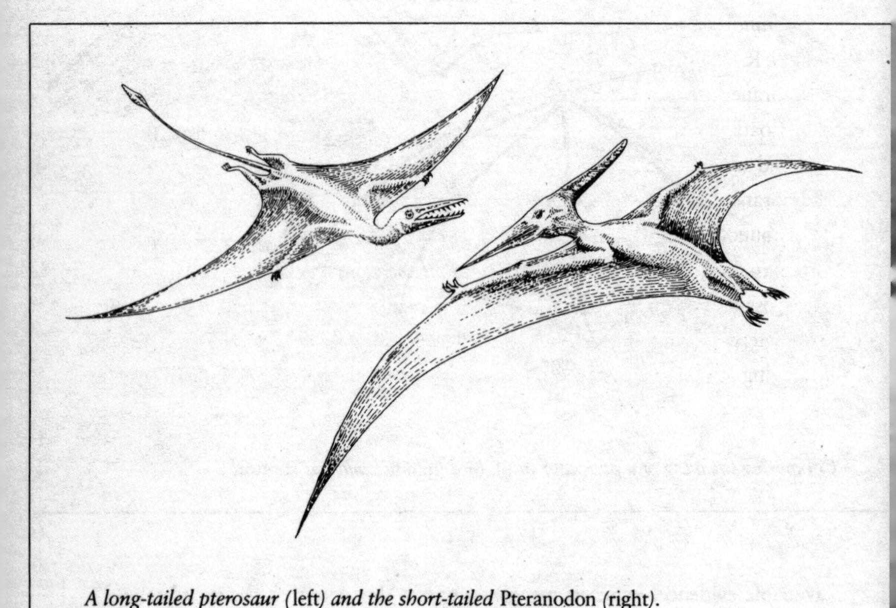

*A long-tailed pterosaur (*left*) and the short-tailed* Pteranodon *(*right*).*

Regardless of controversies over the extent of the wing membrane, and the corollary arguments regarding how well pterosaurs walked, there is complete agreement that they were remarkably lightly built. As with birds and bats, their bones are hollow, but the walls are relatively much thinner and their skeletons are consequently flimsier. The smallest pterosaurs were about the size of a starling and were probably flapping flyers, but the larger ones—pelican-sized and larger—were probably soarers. Many attempts have been made to estimate the flying characteristics of pterosaurs. Because soaring is much simpler to analyze than flapping flight, *Pteranodon*, one of the best preserved of the large pterosaurs, has been the focus of the most attention.

Pteranodon had a wingspan of 23 feet (7 meters), twice that of the largest albatross. While any attempt to deduce its flying characteristics is bound to be speculative, some idea of the likely parameters is possible. The first step is to estimate the body mass and deduce the wing area. In the first major study of *Pteranodon*, a collaboration between Cherrie Bramwell, a paleontologist, and G. R. Whitfield, an aeronautical engineer, it was assumed that the wing membrane extended to the ankles, giving *Pteranodon* a correspondingly low wing loading of 0.75 pounds per square foot (3.6 kilograms per square meter) and an aspect ratio of 10.5. In a later study by another specialist the wing membrane was not extended to the legs, which yielded a higher wing loading (1.2 pounds per square foot or 5.9 kilograms per square meter) and aspect ratio (19.0). Even the highest wing loading is less than half that of the largest albatross and is comparable to that of a herring gull (wingspan 4.7 feet or 1.4 meters) or white-headed vulture (wingspan 7.3 feet or 2.2 meters). This low wing loading is quite remarkable considering the enormous differences

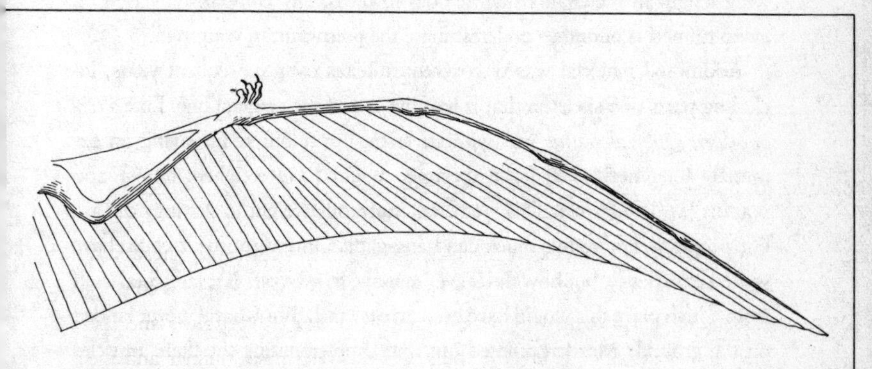

Diagrammatic representation of a pterosaur's wing, showing the orientation of the fibers. For simplicity only about 2 percent of the fibers are represented.

in size. The combination of a low wing loading and high aspect ratio is like that of the frigate bird, which serves as a living analogue for pterosaurs. *Pteranodon* is envisioned as a very maneuverable soarer, capable of generating more than enough muscle power to become airborne, but too large to have been able to sustain flapping flight for any length of time.

Since the time of its discovery, almost a century ago, *Pteranodon* was the largest pterosaur known. Then, in the early 1970s, paleontologists found some gigantic remains in Texas that turned out to be a pterosaur. Named *Quetzalcoatlus* after the feathered Aztec deity, the new specimen was far from complete, comprising part of one wing and a few neck vertebrae, but they were distinctive because of their gigantic size. The humerus, one of the shortest bones in the pterosaurian wing, was almost 2 feet long (52 centimeters), while the neck vertebrae, which are remarkably long and slender, are over 1 foot long (40 centimeters). The initial estimate for the wingspan was 50 feet (15.5 meters), but later work suggested that 36 to 39 feet (11 to 12 meters) was probably a closer estimate—the size of a light twin-engined airplane. One critic argued that the wingspan was a lot smaller, but this was based on a comparison between the length of the humerus of *Quetzalcoatlus* and that of birds. Since the humerus forms a much smaller portion of the wing length in pterosaurs than it does in birds, any comparison based on avian humeri is bound to underestimate the pterosaurian wingspan.

Additional material was discovered in Texas over subsequent years, including parts of a skeleton that is half the size of the original one. Like *Pteranodon*, *Quetzalcoatlus* was without teeth, but it differs in having an extremely long neck. The large specimen might have weighed almost 200 pounds (86 kilograms). This seems an enormous weight to get into the air, but people weighing this much can hang-glide. Once airborne it could have soared effortlessly, but how did it ever manage to take off? If it launched itself from a high place this would have been an easy task. But what if it ever landed on the ground? Marden pointed out that supplementing the flight muscles with the more powerful anaerobic fibers would have provided ample power for takeoff. But I still have some difficulty imagining an animal with a wingspan of almost 40 feet being able to flap its wings fast enough to take off.

This raises the question of the lifestyle of *Quetzalcoatlus*. If it lived by the sea it could have fed on marine organisms, feeding on the wing like many present-day seabirds. However, in marked contrast with almost every other pterosaur that has been found, it was collected from non-marine sediments. It has been suggested that it was a carrion feeder, whose long neck probed deep inside the decomposing bodies of dinosaurs, like some serpentine vulture. The hulking carcass of a sauropod might have provided a suitable launching pad for takeoff too, but this entire scenario strikes me as fanciful. Perhaps the images of those flesh-gorged condors struggling to become airborne are too firmly entrenched in my mind. In any event, I have no alternative suggestions to their possible lifestyle. *Quetzalcoatlus* raises more questions than it answers, but we can be quite sure that it *did* fly even though we might not understand *how*.

TIFFANY WINGS AND KITE STRINGS

Each fall, in school gymnasiums across the land, groups of enthusiasts reconvene after the summer recess to indulge in their passion—flying indoor model airplanes. Not knowing of the existence of this genre of airplane, far less of the dedicated model-makers that fly them, I came upon both by chance. Some time ago I turned on the television to catch the last few moments of a documentary film. I recall nothing of the commentary, but the images that filled the screen were unforgettable. Imagine an armada of strange aircraft sailing like gossamer galleons across a darkened room. Giant propellers barely turning, transparent wings flexing and reflecting the colors of the rainbow. I did not know it at the time, but this was the world of microfilm models, the name of a class of indoor model airplanes flown in competitions around the world. Built to international standards of size, the aim of the competition is to stay aloft for as long as possible; the world record stands at almost fifty minutes. By a circuitous route I made contact with a local group and went along to watch.

One of the first models to catch my eye was a tiny World War I biplane with a wingspan not much wider than my hand. This was not one of the special microfilm airplanes; it was made of tissue paper and balsa wood like the ones we built in our youth. After winding up the rubber-band motor with a tool like a regular hand drill (all models are rubber powered) and checking the tension with a torque meter (flying is taken seriously here), the owner set his model down on the floor and let it go. The beautiful little airplane taxied across the wooden floor, made a perfect takeoff, and proceeded to climb in well-behaved circles. After a few minutes it reached its maximum height; then, as the power began to taper off, it began a gently spiraling descent. The flight ended with a landing that would have made any pilot proud.

I saw other models fly similar routines, but I was anxious to see some microfilm models. Then I learned that none were there that night, and for good reason. The conventional models, being much heavier and flying at higher speeds, could tolerate the disturbed air of the gymnasium. Microfilm models, though, required still air, an impossibility with people moving about. I got the telephone number of a man who built and flew microfilm models who might help me.

Mike Thomas, a retired engineer, met me at the appointed time at a school gymnasium on the other side of town. By a long-standing arrangement with the school janitor, the heating system had been turned off to minimize air disturbances. We had the place to ourselves, which suited Mike well, and he set to work preparing his microfilm model for the flight. He transports the airplane in sections, in a large custom-built box that looks like a small workshop, with everything neatly stowed and in its right place. I was prepared to see a lightweight airplane, but was amazed by this one's extreme frailty. The wing, tail, and propeller were built of the flimsiest threads of balsa wood covered with a plastic film as thin as a soap bubble. Mike handed me a postage-size scrap of film from his repair kit. I touched it as gently as I would a delicate fossil; it crumbled to dust between my fingers. He smiled and remarked that

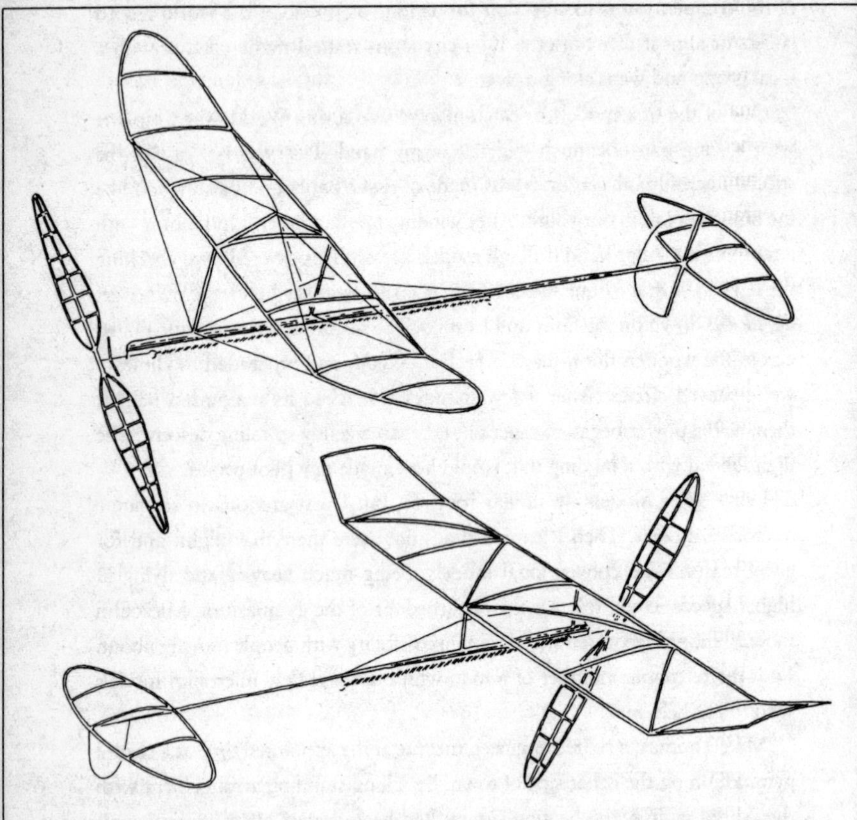

These microfilm model airplanes have wingspans of about 2 feet (65 centimeters) and weigh less than 3 grams.

I should try building an airplane out of it! He explained that the film is made by pouring a liquid on the surface of a bath of water. The next trick is trying to beat the forces of surface tension to lift it off the water. Often as not the film breaks, or it is not quite big enough for what you want, or it breaks when you try laying it down over the airframe. This is not a hobby for the impatient. The fuselage is the heaviest and most rugged part of the aircraft because it has to withstand the compressive forces of the rubber band. It is made of balsa wood only 0.3 millimeter thick, rolled into an 8-millimeter-diameter tube reinforced with fine filaments of boron. The model belongs to a class designated F1D by the F.A.D. (Fédération Aeronautiqué Internationale), specified as having a maximum wingspan of 65 centimeters (25.5 inches) and a minimum weight (less the rubber band) of 1 gram. Mike's airplane weighed 1.2 grams, 2.7 grams with the rubber motor—roughly the weight of a grape.

It flew like a living creature, rippling and flexing to the thrust of the propeller and the flow of air over its delicate skin. To get the best possible view I kept about a yard in front of the aircraft, walking slowly backward while it flew toward me. I had to be very careful not to create any undue turbulence; a slight wave of the hand within a couple of feet would have set it in turmoil. In spite of my precautions I managed to create a small patch of turbulent air in its path that caused it to stall and tumble toward the ground. But my companion saved it so no harm was done.

We measured its speed, which varied between 0.52 and 0.65 meters per second (1.2 to 1.4 miles per hour). Taking the lower end of the speed range gives a Reynolds number of about 5000, which is lower than that of bats and birds and is in the range of large insects like the hawk moth. While this is not an especially low Reynolds number, it is sufficiently low that air flows smoothly over surfaces in an orderly fashion. Wind tunnel experiments have shown that, even at angles of attack as high as 30 degrees, the airflow remains in contact with as much as about 85 percent of the chord (width of the wing). As in insects, the wing is a simple, thin, curved plate, quite unlike the conventional wing of birds and airplanes where an appreciable thickness separates the top and bottom surfaces. The significance of this is that conventional airfoils are less efficient than simple curved plates at a low Reynolds number because boundary-layer separation takes place. Because of its ultralightweight construction the wing loading is remarkably low—only about 0.02 kilogram per square meter (0.07 ounce per square foot)—which is about 1/130 that of a small bird and 1/18 that of a fruit fly. This small wing loading makes the model so sensitive to air movements.

Flight first evolved over 290 million years ago, during the Carboniferous

Period, with the appearance of flying insects, including dragonflies. Like our own early attempts at flight, these insects had two pairs of wings, but these were set one pair behind the other, not stacked as with biplanes. Some groups, like the Diptera (meaning "two wings" and including flies and mosquitoes), have undergone reduction to a single pair. Most insects have retained the more primitive double pair of wings, but these are often linked together along their edges so that they function as a single pair. Certain moths and beetles are larger than the smallest bats and birds, but most flying insects are small and many of them are minute. Reynolds numbers for insects therefore range from several thousand to less than twenty, ample material for an investigation of flight dominated by viscous forces.

Most insects have thin membranous wings that are usually transparent and are stiffened by veins, like a leaf. Insect wings are usually uncambered at rest, but when they start beating the air pressure gives them a cambered pro-

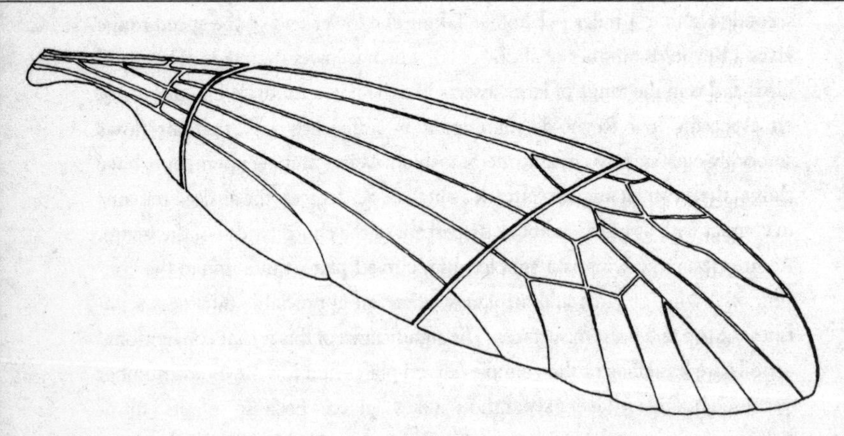

An insect's thin, membranous wing is stiffened by veins. When at rest the wing is flat, but during flight the air pressure gives it a cambered profile. The diagonal arrangement of the veins causes the wing to twist so that the camber decreases toward the tip, as in an airplane's propeller.

file. Roland Ennos, an English biologist at the University of York, investigated how this occurs and discovered a mechanism that was elegantly simple. The leading edge of the wing is stiffened by a main spar, from which other veins branch off at intervals, running diagonally outward and backward across the width of the wing. This diagonal arrangement causes the entire wing to twist when acted upon by air pressure, such that the camber decreases from the base toward the tip. The beating wing therefore takes on the shape of an airplane's propeller, the pitch (amount of twist) being greatest at the root, where airspeeds are least, and least at the tip, where airspeeds are highest. This maximizes the efficiency of the device—propeller or wing—by generating the same amount of lift along its length regardless of local speeds. Ennos also pointed out that a similar diagonal arrangement occurs in the attachment of the flight feathers to the wing skeleton of a bird, and in the oblique arrangement of fibers in a pterosaur's wing.

Like the rest of the insect's skeleton, the wing is made of chitin, a tough material belonging to the same chemical group as cellulose, the structural material of wood. Beetles can bend and fold their wings away somewhat like bats and birds, but most insects simply lay them flat against the body when not in use. Insects, like all other arthropods (jointed-legged animals), have their skeletons outside their bodies (called the exoskeleton). This chitinous casing is stiff but thin and springy, like an unopened can of beer. The wings are hinged to this chitinous capsule in such a way that their up and down movements are aided by the skeleton's natural resilience. Most insects' flight muscles do not even attach to the wings; instead, their movements are brought about by the up and down motion of the body capsule. These arrangements permit strain energy to be stored in the body capsule, and much of the energy released by the flight muscles during the downstroke contributes to the upstroke, and vice-versa. Compared with the vertebrate system this is far more efficient.

Since insect wings are stiff and cannot be folded for the upstroke, as they can in birds, the wing's angle of attack has to be changed with each stroke; that is, it has to be rotated. The easiest way to visualize the wing action is to stretch your arm at your side. Raise your arm above your shoulder, hand flat,

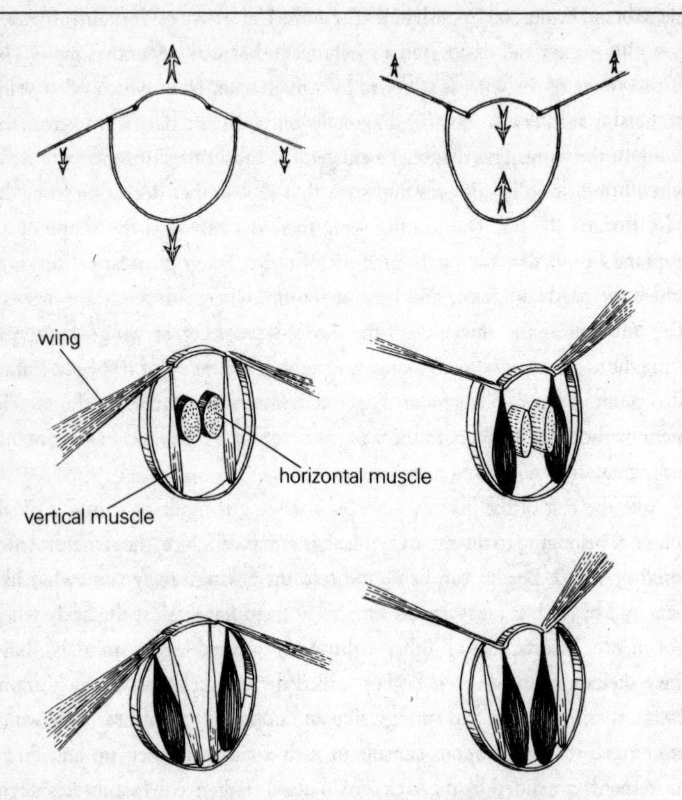

Wing movement in most insects is caused by the up-and-down movement of the springy outer casing of the body, called the exoskeleton, to which the wings are hinged (top). *These movements are brought about by two sets of muscles: a vertical set attaches to the top and bottom of the body, and a horizontal set runs along the body. When the horizontal muscles contract* (middle left) *the body shortens slightly and becomes taller, depressing the wings. When the vertical muscles contract* (middle right) *the body height becomes shorter, raising the wings. In some insects the muscles attach directly to the wings* (bottom). *When the outermost muscles contract the wings are depressed* (bottom left). *Contraction of the innermost muscles raises the wings* (bottom right).

with the thumb side lowered so the palm faces obliquely backward. As you lower and swing your arm forward at the start of the downstroke, your hand functions as an inclined airfoil set at an acute angle of attack; it generates a forward and upward lift force perpendicular to the direction of motion. At the bottom of the downstroke rotate your hand so the upstroke is made with the thumb side raised, palm pointing obliquely up and forward. As your arm cuts through the air, upward and backward, it generates another lift force forward and upward. Each stroke is therefore a power stroke that creates lift and forward thrust. The action is similar to that of hummingbirds, which also have stiff wings that must be rotated after each stroke. Like hummingbirds, most insects can hover.

Decreasing body size is accompanied by four predictable changes: wing-loadings become lower, flying speeds become lower, viscous drag forces be-

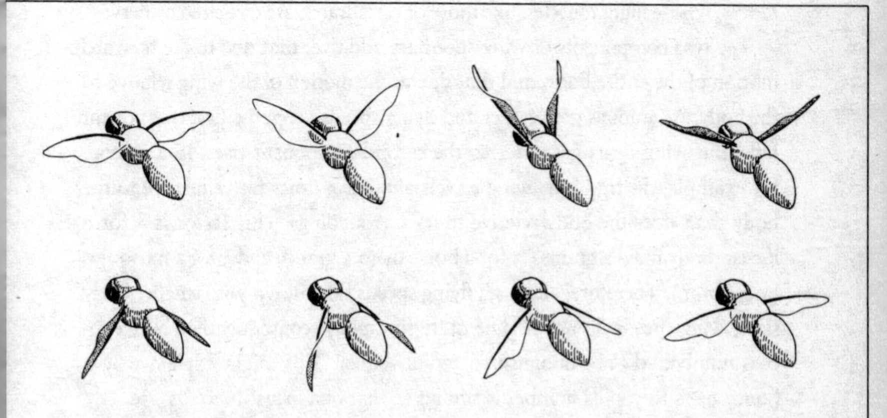

Wing movements of a flying insect. The leading edge of the wing is shown by the heavy line; the underside is shaded. The sequence begins with the wings midway through the upstroke (top left)*. During the upstroke the leading edge of the wing is directed upward, but at the end of the upstroke the wing rotates so that it faces downward* (top, third from left)*. At the end of the downstroke the wing rotates again so that the leading edge faces upward* (bottom, second from left)*.*

come larger, and wing-beat frequencies become higher. One of the consequences of the higher frequencies is that flying insects become audible. Butterflies and moths, being rather large, flap their wings fairly slowly—around twenty beats per second and less—so we cannot hear them. Yet we have no difficulty with the low-pitched drone of a bumblebee (about 150 beats per second) or the annoying buzz of a mosquito (about 250 beats per second). Such rapid wing beats require muscle-contraction rates far higher than vertebrates can achieve, and some insects even have wing-beat frequencies approaching 1000 per second. Vertebrates are unable to achieve such high frequencies because their muscle contractions are controlled by nervous impulses, which cannot occur much faster than about 100 times a second. The flight muscles of most insects, in contrast, are stimulated to contract by mechanical changes in the muscle itself. These changes can occur very much faster—up to 2000 times a second. Among the exceptions are butterflies and moths, whose flight muscles, like those of vertebrates, are triggered by nerves.

The two components of wing speed are additive: that due to the forward motion of the entire body, and that due to the motion of the wing relative to the body. As animals get smaller and flying speeds drop, the first component falls. But wing beats get faster so the second component rises. In a pigeon, for example, the tip of the wing travels about 1.3 times faster relative to the body than does the body relative to its surroundings. This factor is 2 for a locust (body mass 2 grams), 3 for a horse fly (0.2 gram), and 5 for a mosquito (.004 gram). Therefore, although flying speeds slow down with smaller body sizes, wing speeds do not decline quite so rapidly; consequently their Reynolds numbers do not diminish so rapidly either. This might explain why a honey bee's Reynolds number is similar to that of a locust twice its size.

The greater range in wing shape and size among insects than among birds and bats implies more varied flying habits and capabilities. All insects flap their wings, but larger ones, like butterflies and moths, which have particularly large wings, have a low enough friction drag to be able to soar, too. Some butterflies undertake long migratory flights as do birds. Monarch butterflies, for instance, pupate in southern Canada toward the end of the summer and postpone breeding until they have migrated south to winter in Mexico. Many colorful accounts have been penned of migrating butterflies teeming in their thousands through mountain passes, or soaring, en masse, along rugged coastlines to reach their destination. Thousands perish along the way, many falling victim to insectivorous birds and bats, but a sizeable portion always manages to get through.

Butterflies and moths have two primary defenses against aerial predation—some taste terrible and others fly evasively. Species that rely on their flying abilities have more massive flight muscles relative to their body mass than do the unpalatable species. This gives them better acceleration, allowing them to execute erratic zigzagging escape maneuvers. Although avian predators can fly faster in a straight line, they cannot change direction as rapidly as their prey, so the insects often escape. Certain palatable butterflies have evolved external appearances that mimic the unpalatable species, just as certain harmless insects have evolved the warning colors of venomous species like wasps. The mimics of the unpalatable species are avoided by birds, and, they, too, have smaller flight muscles than other palatable species. Butterflies and moths hold much interest for entomologists, professionals and amateurs alike, but from the perspective of aerodynamics insects are more interesting the smaller they get.

To generate lift a wing has to have an airflow over its surfaces, and this can be achieved by moving the wing relative to the body, as in flapping (with or without moving the body relative to the air), or by moving the animal through the air, as in gliding and soaring. As body size diminishes soaring and gliding give way to flapping—and flapping rates speed up as flying speeds slow down. As Reynolds numbers fall, inertial forces decline and viscous forces rise. Lift is an inertial force and consequently drops with declining Reynolds numbers. So too does pressure drag, but friction drag, which is attributable to viscosity, skyrockets. The net result of all this is that as Reynolds numbers decline the lift-to-drag ratio of the wings falls. Eventually the wing can no longer generate sufficient lift to keep the animal airborne. But tiny fliers like *Trichogramma*, which has a wingspan of only about 1 millimeter, still fly, in spite of the burgeoning drag and small lift on their bodies. How do they do it?

The late Torkel Weis-Fogh, an outstanding experimentalist and theoretician of animal flight, set out to resolve this problem. The subject of his attention was a parasitic wasp called *Encarsia*. Similar in size and appearance to *Trichogramma*, it has a similar hovering flight, with two pairs of coupled wings beating in unison. Using a high-speed camera and infinite patience, Weis-Fogh obtained movie footage showing details of *Encarsia*'s wing movements. And there, captured on film, was a brief sequence of wing movements that had been seen by others but whose significance had not been realized. Like other hoverers *Encarsia* flies with its body approximately vertical, the wings sweeping back and forth in a horizontal plane. The wings beat about 400 times a second, at Reynolds numbers of less than 20. The novel feature captured on film was that at the end of the upstroke the two pairs of wings were clapped together, like opposite pages of a book being slammed closed. For an instant the upper surfaces of the paired wings are pressed tightly together; then they are flung apart like a book being opened. The book analogy is a precise one because the top edges of the wing pairs separate first and the bottom edges are the last to part. This rapid opening creates a zone of reduced pressure between the wings. The inrush of air sets up a lift-producing vortex

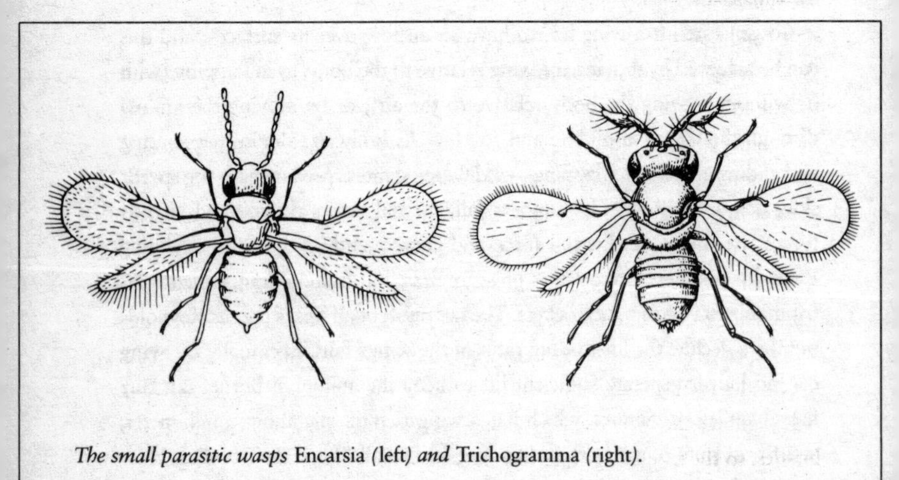

The small parasitic wasps Encarsia *(left) and* Trichogramma *(right).*

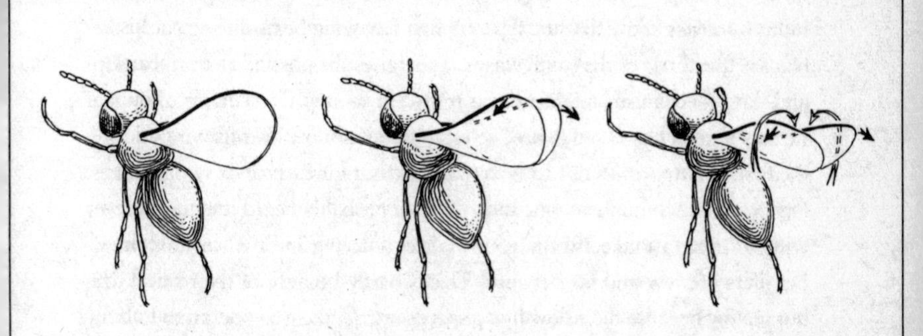

The clap-and-fling flying mechanism of very small fliers like Encarsia *and* Trichogramma. *At the end of the upstroke the wings are clapped together (*left*). They are then flung apart, the separation of the wings commencing with the leading edges (*middle*). The inrush of air sets up lift-producing vortices (*right*). For simplicity only one pair of wings is shown.*

around each coupled wing pair and establishes a fluid flow over their surfaces. This mechanism, described as clap-and-fling, explains how tiny fliers like *Encarsia* and *Trichogramma* can generate lift under seemingly impossible conditions.

The wings of these minute insects are fringed with long bristles, an unusual feature for a wing. Weis-Fogh noted that they were especially prominent on the trailing edges of the hind wings and suggested that they might help to seal the bottom edges of the wings during the initial phase of the fling. He thought that the bristles fringing the forewings might delay the formation of wing-tip vortices, thereby reducing drag. He noted that bristles were absent from the wings of a closely related but much larger wasp, supporting his view that bristles are an integral part of the clap-and-fling mechanism. Charles Ellington, a specialist of insect flight at the University of Cambridge, suggested that the bristles merely increase the surface area of the wings. At such low Reynolds numbers the high viscosity would prevent the inflow of air through the bristles, so the fringe functions like the rest of the wing.

It did not escape Weis-Fogh's attention that pigeons often make one or two distinctive claps when making sudden takeoffs from a standing start. Slow-motion movies show that the pigeon's first few wing beats during such take-offs are like those of the small wasps. This raises the possibility that the clap-and-fling mechanism might not be restricted to tiny fliers. After all, a stationary pigeon has as much need to establish an airflow over its wings during the first few strokes as does a hovering *Encarsia* for each of its wing strokes. Pigeons are common enough, and you have probably heard the clap of their wings during fast takeoffs. But your chances of seeing any low Reynolds number fliers are few and far between. This is partly because of their small size, but mainly because their slow flying speeds restrict them to operating in fairly still air, as with microfilm models.

One of the low Reynolds number fliers most likely to be seen is the fruit fly (*Drosophila*), which hovers around fruit bowls during the summer. They operate at Reynolds numbers of about 200, which also happens to be the frequency of their wingbeats, but you cannot hear them because their wings are so small (with a wingspan of about 5 millimeters) that they do not set enough air into motion to be detected by our ears. They spend much of their time hovering, and when they move they fly slowly enough to be caught by a swift hand. Their wing loading is about 0.35 kilogram per square meter (0.77 pound per square foot), roughly half that of a bee. Fruit flies may be tiny, but compared with *Encarsia* and *Trichogramma*, they are about four times bigger, approximately eighty times heavier, and have wing loadings approximately four times higher. (Remember that weight increases with the *cube* of the size increase—fruit flies are not exactly sixty-four times heavier than *Trichogramma* because they are not exactly the same shape.)

Having read so much about *Trichogramma*, I was anxious to see some for myself and find out how they fly. To this end I visited the Department of Forestry of the University of Toronto and was handed a small and seemingly empty glass vial from one of their refrigerators. On very close inspection I spotted dozens of minute white specks moving around slowly on the inside of the vial. As the heat from my hand warmed the glass they began hurrying around like expectant fathers. They are kept in the refrigerator to slow their metabolism—that way they live more than a week, whereas they live only a day or two at room temperatures. A streak of nectar on the inside of the vial provided all the food that the diminutive wasps needed for the rest of their short lives. When I brought my prize back to the Museum I looked at them under the microscope, but was foiled by their fast movements. The vial was a poor observatory for studying their flying because they had barely enough room to take to flight. I transferred them to a large glass flask, but they were still hard to see because of their minute size and pale color. I resolved the problem by darkening the room and illuminating them with a narrow beam of bright light. Like many other insects, they are attracted by light and made a beeline for the widest part of the flask where the beam was directed.

Every so often one of the wasps would launch into a short hop an inch or so in height. This appeared to be a jump rather than a flight, but the duration was too short to be sure. About as frequently, a wasp would set off on a brief spiraling flight lasting a few seconds, reminiscent of a moth's flight around a light bulb. The diameter of the spiral was less than ½ inch (1 centimeter) and its length was up to 5 inches (12 centimeters).

In the wild these tiny wasps spend their time moving about in spruce trees and laying their eggs inside the eggs of other insects. Being so small, they probably spend their entire life in one tree. I imagine that they probably confine their flying to the relatively still air around the tree, not taking off except when it is calm. To do otherwise would be to risk being carried off by a breeze and possibly perishing instead of finding a second home. To test this idea I introduced a narrow rubber tube into the flask that was connected at the other end to an air line. Sure enough, every time I turned on the air supply any wasps that were flying immediately became grounded (probably by being thrown against the side of the flask). The others made no attempt to fly until I turned off the airflow.

When considering the possible role of wing bristles, Weis-Fogh noted that as insects got smaller the bristles became larger and more extensive relative to the membranous part of the wing. Eventually the wing's main part is simply a narrow rod supporting a broad fringe of bristles. It has been suggested that such wings no longer generate lift using inertial forces, as in other fliers, but that they use drag forces instead. The concept that these fluffy wings are

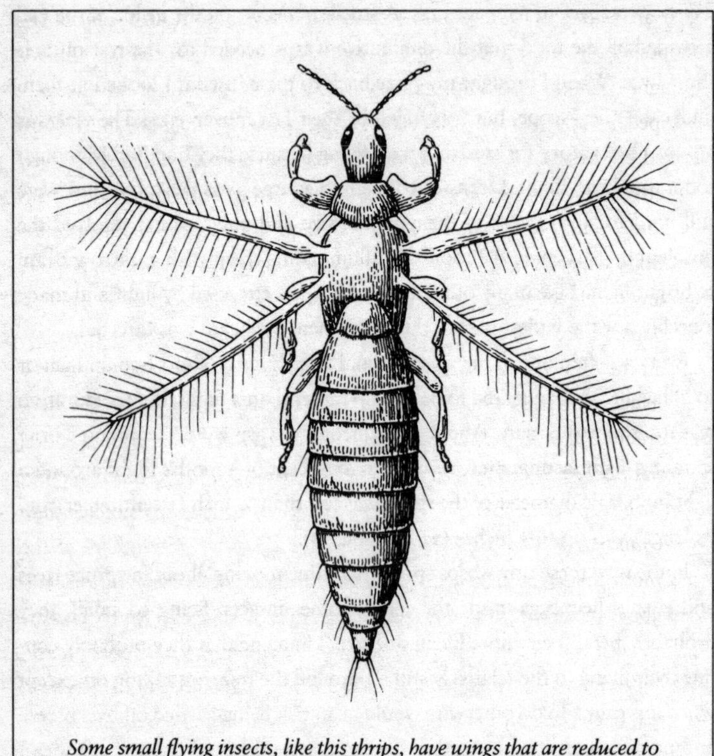

Some small flying insects, like this thrips, have wings that are reduced to rods fringed with bristles. Like the small parasitic moths, they fly using the clap-and-fling mechanism.

used to swim through the air is an attractive one, but like so many other good ideas it is also incorrect. As Ellington showed in his detailed study of insect flight, animals with these sorts of wings have a clap-and-fling mechanism like that of other small insects.

I have not managed to obtain any fluffy-winged insects to look at, but I imagine that their flying powers are on a par with those of *Trichogramma*, and some are probably inferior. How useful are such marginal flying abilities, which are operational only on the calmest of days? From an evolutionary standpoint, the significance of these weak powers is precisely that an individual might be swept away on a chance breeze. Flight, no matter how weakly developed, guarantees the dispersion of the species. The small individuals whisked off by a chance breeze are like the Polynesians who provisioned themselves for a long voyage and steered their flimsy outriggers away from their home island. The chances of being swallowed up by the vast emptiness of the South Pacific far outweighed the chances of finding a new island home, but a few would survive to perpetuate the race.

Though we are usually not aware of it, a vibrant world of living organisms teems in the air around and above us. This aerial plankton is so named because its members, like the more familiar oceanic plankton, freely drift with the currents. Extending to heights of 15,000 feet (4,600 meters) and more, the aerial plankton comprises a myriad of species—primarily insects—ranging in size from large and active fliers like locusts to animals smaller than *Trichogramma*. It also includes non-fliers like the furry caterpillars of the gypsy moth, spiders, mites, seeds, sand grains, and the like. One sample, collected above the Pacific Ocean over 250 miles (400 kilometers) from the nearest land, even contained a bivalve mollusk! Other denizens of the air are microscopic organisms and parts of organisms, including pollen grains, protozoa, bacteria, viruses and DNA molecules—a veritable universe.

Our knowledge of aerial plankton dates back at least to the time of the pharaohs and the arrival of locusts in Egypt on the east wind (Exodus 10:13). Gilbert White, the English naturalist, writing in the latter part of the eighteenth century, recorded a shower of aphids that fell on the village of Selborne and attributed this to a north wind. Captain William Parry, White's fellow countryman, found aphids on ice floes in the Polar Sea more than 100 miles (160 kilometers) from land during his 1827 expedition to the North Pole.

The study of aerial plankton did not begin in earnest until the beginning of the twentieth century, when interest focused on economically important pests like the cotton-boll weevil and the gypsy moth. Many techniques have been employed in sampling: kite strings and balloons fitted with simple sticky traps of cards smeared with grease; nets attached to ships' masts and to the cables of large tethered balloons; and suction traps that filter large volumes of air. Insect traps with fine screens have even been attached to the wings of slow-flying aircraft. From many studies we know that the larger and more powerful fliers are generally found at lower levels, up to about 3,000 feet (920 meters), while the smaller, more buoyant fliers and non-fliers predominate at higher altitudes, up to 15,000 feet (4,600 meters) and higher. English entomologist C. G. Johnson, who has made a major contribution to our knowledge of insect dispersal, warns that this picture may be an oversimplification of reality. Regardless of what the true picture might be, there is no question that insects large and small, together with a profusion of other organisms and all manner of inorganic flotsam, are transported far and wide as temporary members of the aerial plankton. These journeys may be as long as only a few miles or many thousands of miles, sometimes with surprising results.

Many years ago, when I was living in London, I left college late one night to discover that my car was splattered with large brown blotches. I thought at first that this was the work of vandals, until I touched the spots, which felt gritty. The streets showed evidence of an earlier rain shower, but I could not make any connection between the rain and the grit on my car. The mystery was not resolved until the following morning, when I read in the newspaper that a severe sandstorm in the Middle East had injected sand high into the atmosphere and that, during the night, this had rained down on southern England. Egyptian sands on London streets evoke all manner of images, but we have all had less exotic reminders of the presence of aerial plankton. Check out your car's windshields when driving in the countryside on a summer's day. You will find it coated with a myriad of tiny insects.

DRIFTING WITH THE TIDE: LIFE IN THE PLANKTON

I WAS A student in London during the 1960s when field courses were an integral part of the training of zoology undergraduates. In England the sea is never very far away, and since the seashore offers such a diversity of habitats and organisms, our field trips were usually taken on the south coast. In our final year we took a long drive north to a marine biological station on the Northumberland coast, overlooking the North Sea. The average sea depth there is about 150 feet (50 meters); yet sixty miles (100 kilometers) to the southeast—about one-third of the way toward Denmark—the sea bed has shoals as shallow as 50 feet (15 meters), which form the Dogger Bank.

The North Sea has a long history of bad weather. When a heavy sea is running, the powerful up-wellings from deeper waters make the Dogger Bank particularly treacherous for shipping. But the up-welling of bottom waters has the beneficial effect of carrying nutrients into the upper layers, where light levels are high enough for photosynthesis to take place. As a consequence the fertile waters are rich in *phytoplankton*, the microscopic plants of the plankton. The phytoplankton provides abundant food for small animals that drift alongside them—the *zooplankton*—and these, in turn, provide food for fishes. Because of its fertility and before over-fishing sent fish stocks plummeting, the Dogger Bank was one of the richest fishing grounds in the world, and the northeast coast was dotted with fishing communities that sent their boats out to collect the bountiful harvest.

We did not get as far as the Dogger Bank while we were there, but we did spend a fair amount of time at sea, where we did some trawling, bottom-sampling, and towing for plankton. And my most vivid recollection of my undergraduate days was when we collected a night plankton sample. We students had heard of bioluminescence, the phenomenon of certain living organisms emitting light, and we knew that some *plankters* (the term for an individual member of the plankton; *planktont* is also used) were bioluminescent, but we had never experienced the phenomenon ourselves. It was cold and damp standing on the quarterdeck that night, but at least we were not blanketed in a North Sea fog. One of the crew members pointed down at the screw-wash from the propeller. Could we see that faint blue-green glow? Some of us thought we could; others were less sure. If this was bioluminescence it was not very impressive.

The skipper slowed the boat to less than 2 knots and the plankton net was put over the side and towed behind the stern. Plankton nets come in different shapes and sizes, all variations on a theme: a gently tapering cone of fine mesh for straining the plankton, with a glass jar or other suitable container at the far end for collecting it. Although all manner of mesh sizes are available, depending on what sort of plankton is to be collected, two types are generally used: coarse for zooplankton and fine for phytoplankton. Our tow was made with the zooplankton net, and after thirty minutes the net was hauled back on board and the concentrated plankton sample emptied into a collecting jar. If we were ever going to see bioluminescence this was our best chance. The deck lights were too bright so we took the jar into the wheelhouse and turned off all the lights. And there in the dark, as the boat pitched with the idling diesel, the jar scintillated with sapphire light. Those closest to the jar could see that the brilliant blue points of light were organized in rows, each row moving independently of the other ones, like small ships passing in the night. What on earth were they? A flashlight was produced and the small ships took on the translucent form of tiny shrimps, barely an inch long. These crustaceans, called *euphausids*, are known by the common name of krill. The flashlight was switched off and we resumed our *son et lumière* in rapt awe.

The word "plankton" comes from Greek for wandering, an apt description for the myriad of organisms that drift with the currents in the upper layers of lakes and oceans. The depths at which plankton thrive best are determined by light intensity, varying from about 100 feet (30 meters) in equatorial waters, where the sun is more nearly overhead, to about half that depth in high latitudes where the sun's rays do not penetrate so far. Light is clearly essential to phytoplankton, and zooplankton, ultimately dependent on them as a food source, are also limited by light levels in their depth distribution. Most plankters are small; members of the phytoplankton are the smallest, ranging from about 1 millimeter down to something less than the size of a human red blood cell (which has a diameter of 7.5 micrometers, abbreviated µm), and visible only with a microscope. Zooplankters, in contrast, are mainly visible to the unaided eye. Although the larger ones, like the euphausids, reach shrimp sizes, most of them are much smaller. Very large zooplankters do occur though, and there are accounts of gigantic jellyfish in Arctic waters that reach diameters of over a yard (over 1 meter). Size, then, does not determine whether an organism is planktonic. The key characteristic is its swimming power. Some planktonic organisms, like most of the diatoms—which are all phytoplanktonic—have no means of propulsion, while others—especially the bigger ones like euphausids, jellyfish, and fish larvae—have well-developed swimming abilities. Even though these stronger swimmers can move about within the plankton, their limited abilities prevent them from leaving it. Contrast this with most adult fishes, like mackerel and herring, and animals the size of sharks and dolphins. Their strong swimming abilities enable them to travel anywhere at will; they are referred to collectively as the *nekton* (Greek, *nēktos*, swimming).

Many organisms are permanent members of the plankton. They spend their entire lives adrift. Other organisms are only temporary travelers that incorporate a planktonic phase as part of their development. Included among this group are fish larvae and a host of larvae of littoral (seashore) and benthic (bottom-living) animals like barnacles, starfish, limpets, bristleworms, crabs, and sea urchins. These temporary members of the plankton are seasonal, appearing mainly in spring and fall.

Most plankton samples studied in the laboratory have been preserved in formalin, which turns them gray and lifeless—a marked contrast to the teeming world of color seen beneath the microscope when fresh samples are available. The most conspicuous element of the plankton are the copepods, flea-

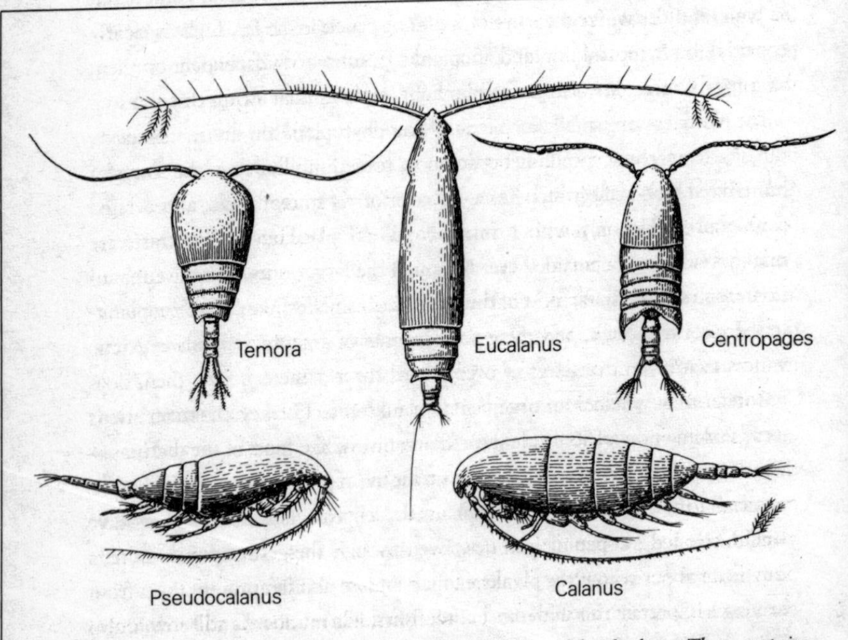

Copepods are small crustaceans that are common members of the plankton. They comprise many different genera and species.

sized (0.5- to 1.5-millimeter) crustaceans with torpedo-shaped bodies and conspicuous antennae (actually *antennules*). Under appropriate lighting conditions copepods are dazzlingly beautiful. Their thin exoskeletons are opalescent, and many of them have a vermilion pigment spot, like a jewel in a crown. They are best seen under the lowest power of the microscope, but smaller members of the plankton require a higher magnification.

While some species of copepods feed on other animals, including other copepods, they are predominantly herbivorous and eat the multitude of phytoplankters that surround them. Like crabs and other crustaceans, copepods have several pairs of small feeding appendages around the mouth, and larger appendages for swimming attach to the main part of the body. By beating their feeding appendages at rates up to 50 hertz, copepods create vortices in the water that bring food particles streaming past the mouth. Some species probably obtain most if not all of their food this way; others, like the species that prey on other crustaceans, select and seize individual food items.

The vortices set up by the feeding current propel the entire copepod through the water at speeds of between 0.5 and 2.0 millimeters per second. This may seem painfully slow, but, relative to the animal's size, it is a reasonable speed. That is, it amounts to approximately one body length per second—equivalent to a porpoise cruising along at about 4.5 miles per hour (7.2 kilometers per hour). In marked contrast to a porpoise, though, the Reynolds number of a cruising copepod is exceedingly low (roughly 1). Consequently its movements are strongly influenced by viscous forces.

Copepods are difficult to observe because every so often one of them gives a few violent jerks of its body and zooms out of the field of view. This is their escape response. In such situations their five pairs of swimming legs propel them forward at speeds as high as 20 centimeters per second. For a 2-millimeter-long copepod like *Calanus* this is a staggering one hundred body lengths per second, equivalent to a porpoise accelerating to 450 miles per hour! Impressive as this may be, the Reynolds number is still low, only about 400. One of the most striking effects of this high viscosity situation is that as soon as the copepod stops swimming its body comes to a halt. A fish can give a few flicks of its tail and then glide for several body lengths, but a

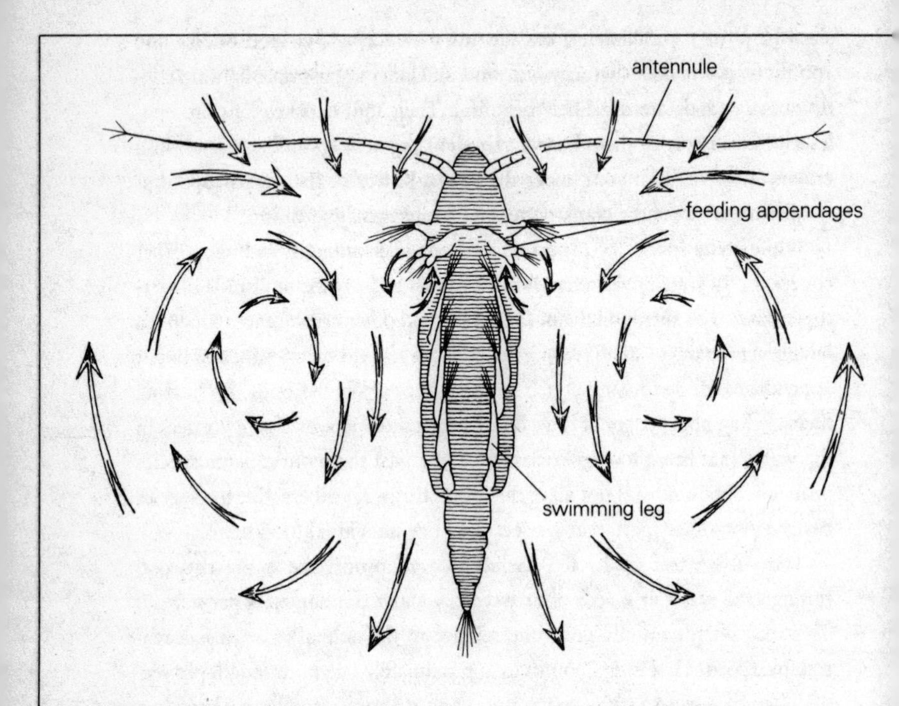

The feeding currents set up in the water by copepods propel them slowly forward.

copepod stops as if it had run into a brick wall. During these sprints the swimming legs beat rapidly back and forth, generating a backward movement of water that propels the animal forward. The Reynolds number of these swimming movements is about 500, which biologist Rudy Strickler of The Johns Hopkins University, Baltimore, considers to be at about the lower limit for inertial propulsion.

Copepods have voracious appetites and can devour all of the phytoplankters in their immediate vicinity—like graduate students at a buffet. During such bacchanalian frenzies much of the food passes straight through their gut undigested, and copepods can so decimate the phytoplankton as to bring about their own demise through starvation. But these feeding frenzies may be exceptional and perhaps most of their time is spent observing more polite table manners.

Next to copepods, the most common organism in a plankton sample are diatoms. They come in an overwhelming array of shapes and sizes. Although many species remain as isolated cells, others join together to form chains and clumps. The smallest cells are a little larger than a human red blood cell, but the largest ones are about one-tenth of a millimeter across. All diatoms have a siliceous (glassy) exoskeleton that comprises two halves, one half fitting snugly inside the other, like the two halves of a petri dish. Many of them,

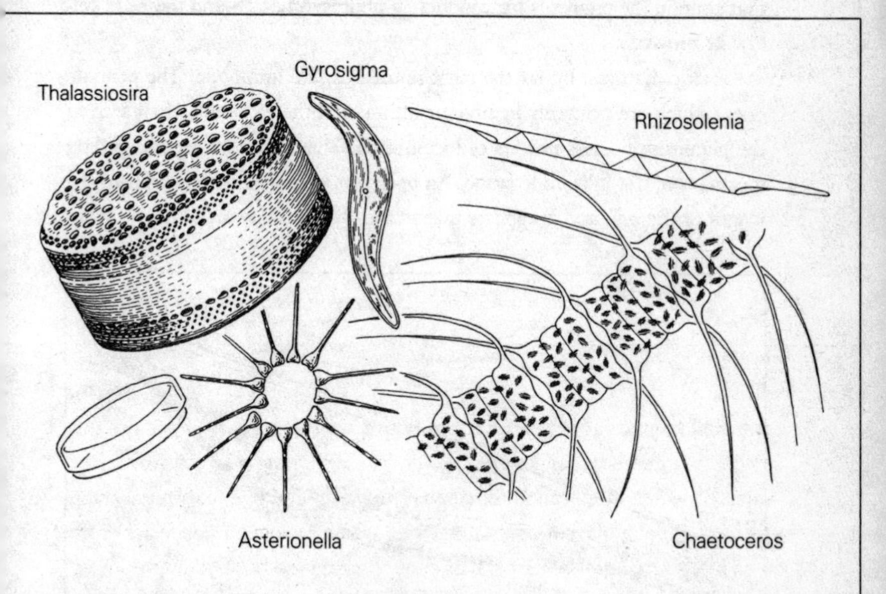

*Diatoms come in a rich array of shapes and sizes. Their glassy exoskeletons comprise two parts, like the two halves of a petri dish (*bottom left*). Many are solitary, like the three at the top:* Thalassiosira, Gyrosigma, *and* Rhizosolenia. *Many others form colonial chains. These might be straight, as in* Chaetoceros, *or coiled as in* Asterionella.

especially the colonial ones, possess elongate processes that increase the drag forces on them. Hence they tend to stay where they are in the water column. Their density is often slightly more than that of the surrounding water, and so they tend to sink, albeit slowly because of drag forces. If a diatom's skeleton is examined under a high-power microscope, one finds that its surface is finely sculptured. One group, called centric diatoms, has patterns that radiate out from one or several points. Others have sculpture lines that are rather straight. This feathery appearance gave rise to their name of pennate diatoms. Centric diatoms take various shapes: discs, pill-boxes, lozenges, flat ribbons. But the pennate ones remind me of vintage battleships because of their boatlike shape. Being plants, the diatoms have chloroplasts—bodies that contain the pigments for conducting photosynthesis—and these are yellow or brown.

Centric diatoms, by far the most numerous, are immobile. The pennate ones, which are primarily benthic or littoral and therefore not often seen in the plankton, possess powers of locomotion. They move by an odd gliding mechanism, not fully understood. An open groove, called the *raphe*, runs the length of the cell, and it appears that strands of mucilage are extruded from

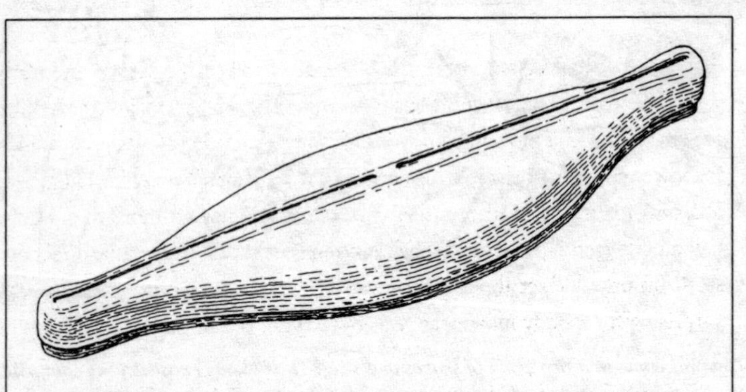

*A pennate diatom (*Navicula*). Notice the longitudinal groove called the* raphe. *Some diatoms, including this one, have a raphe at the top and bottom.*

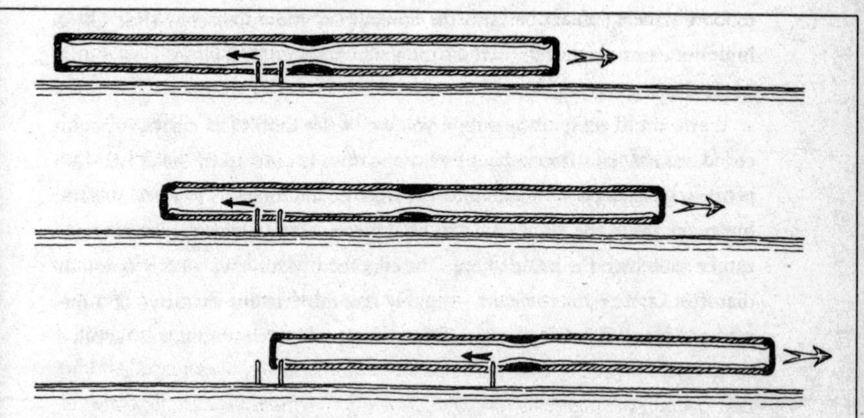

Motility occurs only in pennate diatoms and involves a jerky gliding motion. A longitudinal section through a diatom like Navicula *shows two threads of mucilage being extruded through the posterior half of the lower raphe* (top)*. The tops of the threads are embedded in a specialized layer of cytoplasm that moves rearward, like a conveyor belt. Since the ends of the threads are embedded in the substratum, the whole cell moves forward, like a gondola* (middle)*. New threads of mucilage form to replace the ones left behind on the substratum* (bottom)*.*

the raphe's center region. These strands harden by absorbing water and form stilts that make contact with the substratum. The upper ends of these stilts are embedded in a moving conveyor belt of specialized cytoplasm, and since the lower ends are anchored, the entire cell is propelled along like a gondola. Diatoms move from 1 to 25 micrometers per second, no more than about half a body length per second. Their motion is jerky; they go in one direction, stop abruptly, and set about in the reverse direction. Such movements are not surprising for a body moving at very low Reynolds numbers in the region of 10^{-4}.

Centric diatoms, with their siliceous spindles and cut-glass ornamentation, are elegantly attractive. As you sit and admire their quiet beauty something zooms into view and streaks out again just as quickly. By reducing the magnification to increase the field of view you can locate the speeding protozoan, dashing about beneath the coverslip of the microscope slide like a bull in a china shop. You increase the magnification for a closer view, but it keeps moving too fast.

If you could get its cooperation you would see that it is a ciliate, a single-celled animal (like *Paramecium*) whose surface is covered by small hair-like processes, called cilia. Ciliates measure from 20 micrometers to 2 millimeters, but most are in the range of 50 to 500 micrometers; hence the largest ones can be seen with the unaided eye. The cilia themselves have a fairly constant diameter of 0.25 micrometers—roughly one-thirtieth the diameter of a human red blood cell. Cilia have a uniform internal structure that was revealed

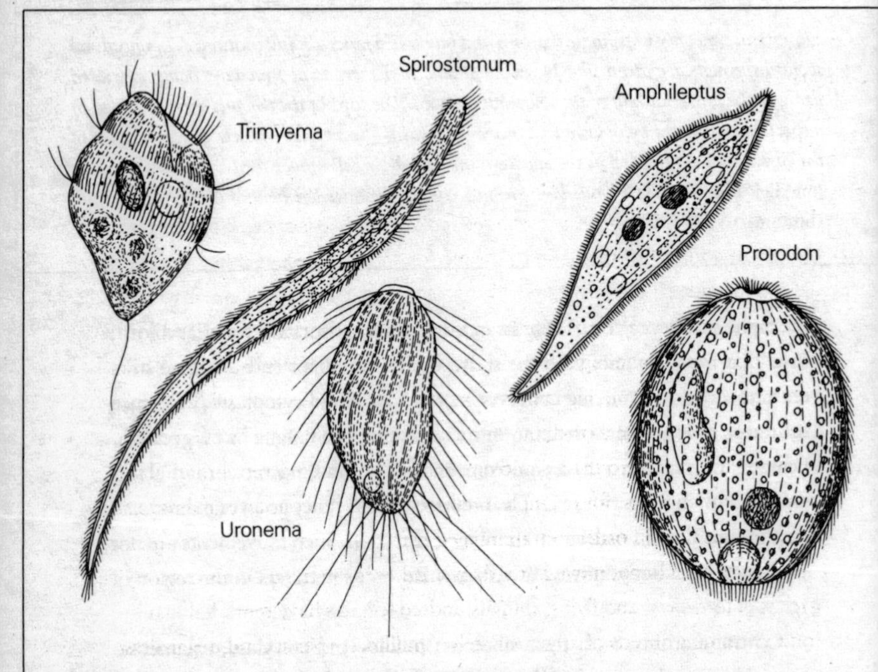

Ciliates move rapidly, propelled by numerous hair-like processes called cilia. *The smallest ciliates are not much larger than human red blood cells but the largest ones are up to 2 millimeters long.*

only after the invention of the electron microscope. Their cross sections show that two central filaments are surrounded by an outer ring of nine paired filaments. The cilium moves by building cross bridges between adjacent members of the outer filaments, similar to the way in which muscles contract. The making and breaking of the cross bridges cause the cilium to bend. The cilia beat in waves like a field of wheat being ruffled by wind. During the power

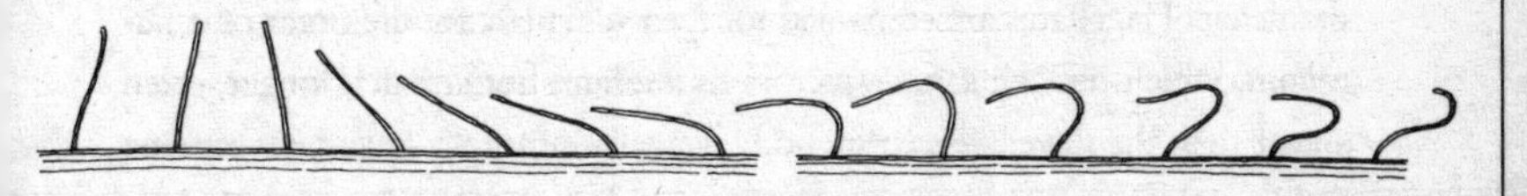

During the power stroke the cilium is kept stiff (left), *but the recovery stroke is made with a limp cilium* (right).

stroke each cilium is stiff; it acts like an oar against the water's drag and produces thrust. When a boat is rowed the recovery stroke is made with the oar out of the water, but this is not possible for the cilia because they are immersed in water. If the recovery stroke were made with a stiff cilium it would negate the propulsive thrust of the power stroke, no matter how slowly it moved, because of the high viscosity (Reynolds number of the cilium is between 10^{-1} and 10^{-2}). The recovery stroke is therefore made with a limp cilium that keeps close to the body surface to minimize drag. The cilia found in ciliates are similar to those in other organisms, including vertebrates. The lining of human lungs, for example, is ciliated so the mucous layer maintains a continuous flow in order to trap inhaled particles.

Large ciliates do not move faster than small ones—a trend that is contrary to that of terrestrial and flying animals. Indeed, ciliates have remarkably uniform swimming speeds of approximately 1 millimeter per second, regardless of size. Therefore, if speed is judged relative to body size, the larger ones swim relatively slower than do the small ones. A few ciliates can swim at much higher speeds for short bursts and this is achieved by the use of bundles of extra long cilia that beat very rapidly. Ciliates operate at Reynolds numbers between 1 and 10^{-2}, and if their incessantly beating cilia ever stopped, they would come to a sudden halt. Trying to observe them is a frustrating business. One option is to paint a ring of grease or nail polish around the cover slip to exclude the air, then wait until they run out of oxygen.

Far more cooperative than the ciliates are the flagellates, which travel at about one-tenth their speed—at a rather uniform rate of 0.1 millimeter per second, regardless of size. On average they are one-tenth the size of ciliates (1 to 50 micrometers), though they overlap with the lower end of the ciliates' size range. Flagellates are so named for their whip-like motile organ, the *flagellum*, which has the same structure as a cilium but is much longer, often longer than the flagellate's entire body. Flagellates usually have between one and four flagella. They undulate these to produce thrust, either in one plane or in a helical fashion. Sperms also have flagella and are a convenient means

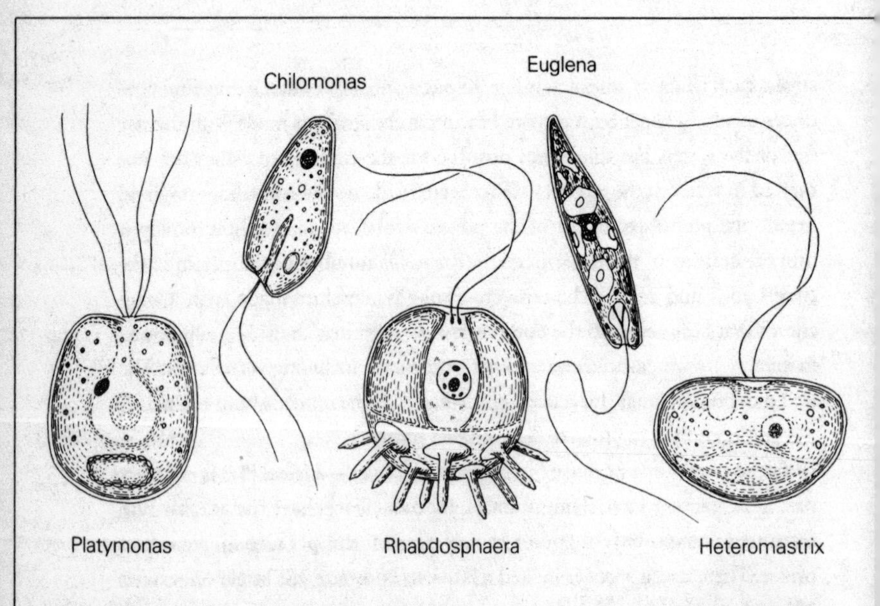

Flagellates are microscopic, ranging in size from 1 to 50 micrometers, and they move about one-tenth more slowly than ciliates.

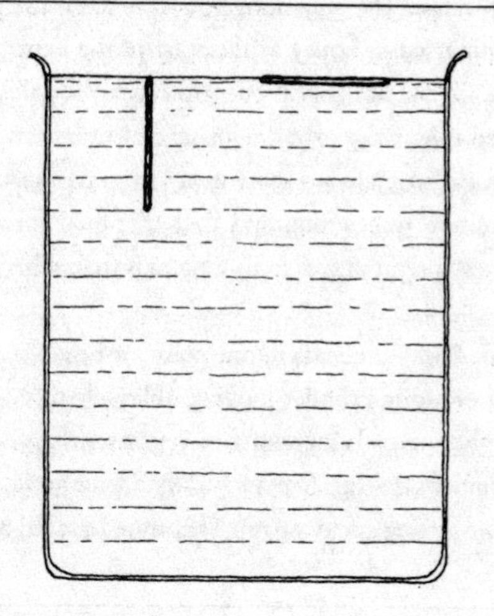

Two equal and straight lengths of paper clip are allowed to sink in a glass of honey or glycerine, one vertical and the other horizontal. The vertical one always sinks fastest.

of studying flagellar structure and movement. While the undulating flagellum of a sperm or a flagellate looks like an eel's undulating body, the resemblance is merely superficial because they operate on entirely different principles. The eel's body works at high Reynolds numbers and generates thrust through inertial forces, whereas the flagellum operates at very low Reynolds numbers, and creates thrust by exploiting drag forces. The underlying principle of the flagellum is that drag forces acting at right angles to its long axis are higher than those acting parallel to it, and these differences are exploited to generate thrust.

I must confess that this mechanism strikes me as counter-intuitive because the predominant drag force at low Reynolds numbers is friction drag, attributable to viscosity, which acts *parallel* to the direction of flow. The drag force acting parallel to the flagellum's long axis should, therefore, be greater than the one acting at right angles. To see for myself I carried out a simple experiment with a tumbler of honey and a paper-clip, which you might like to try. (Glycerine is better than honey because it is clearer, but I did not have any handy.) Straighten out the paper-clip, break off two 1-inch lengths (2.5 centimeters) and let them sink in the honey, one vertically and the other horizontally, to see which sinks fastest. The only thing you have to ensure is that both test pieces are fully immersed in honey at the start of the experiment. This is especially important for the horizontal one; otherwise, it will simply sit on the surface. I repeated this sticky experiment several times, and each time the top end of the vertical wire always drew away from the horizontal wire, showing that my intuition was wrong and that frictional forces *are* higher when the object moves at right angles to its long axis than when moving parallel to it.

Regardless of whether the flagellum beats in one plane or helically, its action can be reduced to that of a thin cylinder moving obliquely through the water. The drag force at right angles to its long axis is greater than the one parallel to it, and so the cylinder side-slips like a log sliding lengthwise down a hill. I demonstrated this by getting sticky again. This time I placed a sheet

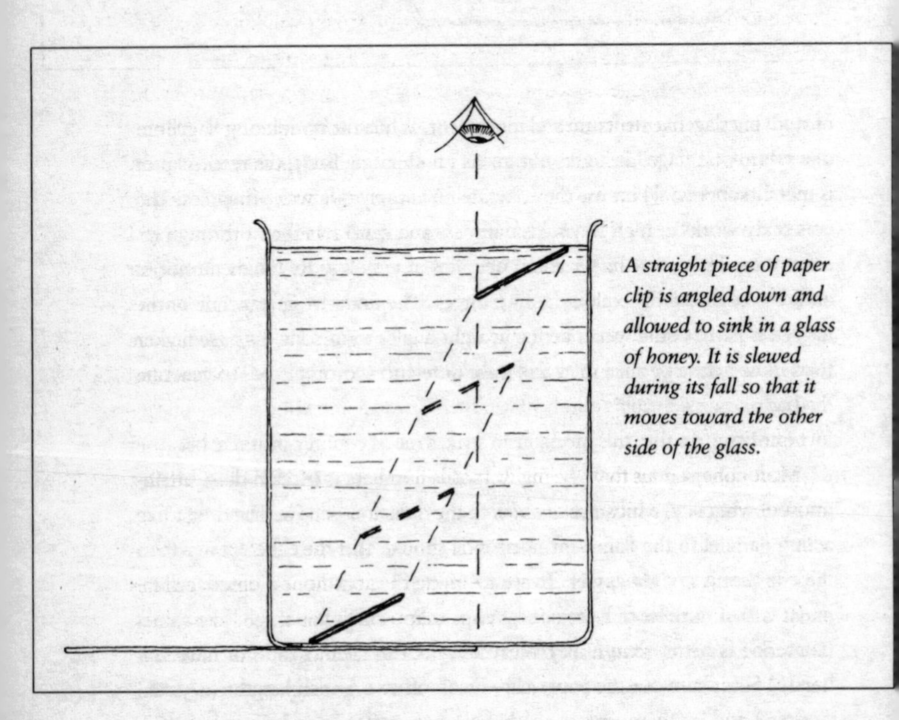

A straight piece of paper clip is angled down and allowed to sink in a glass of honey. It is slewed during its fall so that it moves toward the other side of the glass.

of paper beneath the tumbler of honey. Then I put one of the lengths of paper-clip in the honey, close to, but not touching, one edge of the tumbler. The other end was pointing obliquely down. Before letting go I closed one eye and lined up my other eye with the depressed end of the wire. I adjusted the paper so one edge lined up with my eye to give me a starting line. I let go of the wire and saw that the wire slewed toward the other side of the tumbler, as predicted. The undulations of a flagellum make it act like an inclined cylinder moving through water and generating a propulsive thrust toward the depressed end. A similar mechanism operates for cilia.

As you peer through a microscope, like an explorer from another world, you might see a flagellate swim into view. The beating of its long flagella make the body wobble slightly; and because the water has the consistency of syrup at these low Reynolds numbers, this movement is transferred to the surrounding water. (If you did the honey experiment you will know that it is impossible to disturb a viscous liquid without creating movements at a distance too.) The passage of each organism therefore leaves a signature in the water. These movements might be detected and used by some of the larger predators that move at higher Reynolds numbers to lead them to their prey. Calanoid copepods, for example, leave a less distinctive wake that dissipates about ten times faster than that of their shorter-lived relatives, the cyclopoid copepods. Thus there might have been a selection pressure for the evolution of a less conspicuous wake, just as there is a premium in nuclear submarine design for screws that generate a quieter wash. Generating a less conspicuous wake has potential survival value only in relation to other plankters; fishes and other high Reynolds number swimmers are deaf and blind to the nuances of viscosity.

More conspicuous than the highly motile flagellates are the dinoflagellates, most of which have substantial outer casings of cellulose. They have two flagella; one lying in a prominent horizontal groove and the other in a vertical groove that is not always as obvious. Among the commonest dinoflagellates is *Ceratium*, a spidery looking organism that usually has three long spines. Some species of *Ceratium* are bioluminescent, but the most spectacular emitter of light is *Noctiluca*, a huge gelatinous sphere that reaches at least 1 mil-

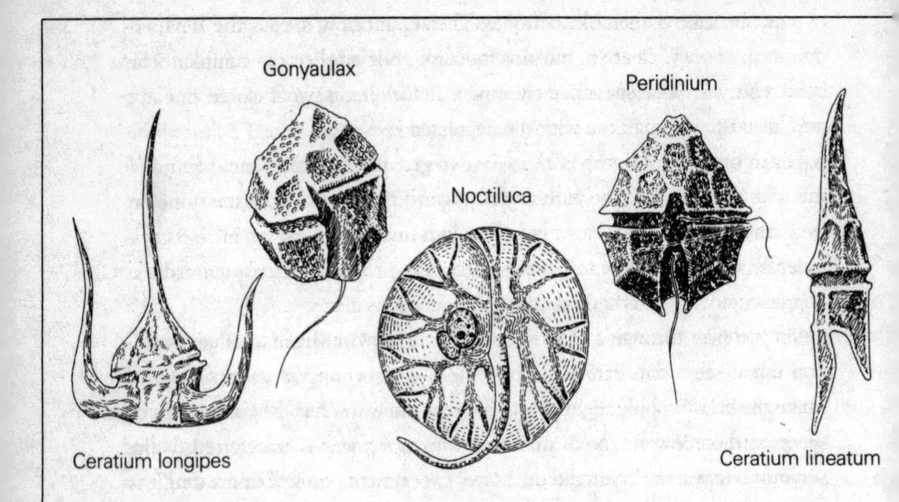

Dinoflagellates have two flagella that lie in separate grooves, the horizontal one being the most prominent. Some are microscopic but others range in size up to about 1 millimeter.

limeter in diameter. *Noctiluca* often occurs in dense swarms, especially near coasts, and is the primary organism responsible for bioluminescence in temperate waters. Much smaller, but far more important, is *Gonyaulax*, the organism responsible for paralytic shellfish poisoning. This flask-shaped dinoflagellate, about 75 micrometers long, often swarms in such dense concentrations that they color the sea red, creating the phenomenon of red tides.

The smallest planktonic organisms are bacteria, which are also the most numerous (up to 1 million per milliliter). Smaller than human red blood cells, they range in length between 0.2 and 5 micrometers and are usually cylindrical. They are motile and propel themselves by flagella, but these flagella are structurally and functionally different from all others. The most interesting feature of bacterial flagella is that they rotate. Indeed the joint at the base of this flagellum is the only example of a rotating joint in the entire living world. Because of their extremely small size and slow speeds the Reynolds numbers of bacteria are in the vicinity of 10^{-5}. It has been calculated that if a bacterium suddenly stopped rotating its flagellum, it would come to a stop in a distance less than the diameter of a hydrogen atom.

Bacterial locomotion has been the subject of considerable research over the last few decades. One of the most intensively studied organisms is *Escherichia coli*, a harmless bacterium that lives in the human gut. It is so frequently used in research that it has become the bacteriologists' equivalent of the laboratory rat. Cylindrical in shape, *E. coli* is 2 micrometers long and 1 micrometer in diameter, with about six long slender flagella. Its swimming movements are quite erratic; it moves in one direction for a short distance, then moves off in another one. Close inspection, however, reveals a pattern. For about one second the bacterium moves in a roughly straight line, called the run. During the run the flagella work together in a bundle, rotating in a counterclockwise direction, propelling the body in front of them. The organism cannot keep on a strictly straight path during a run because of a phenomenon known as Brownian movement. Water molecules are in perpetual random motion and are continually bombarding the particles they surround. If the particles are very small, like bacteria, they are jarred by these collisions, which cause them to jiggle randomly. So it does not matter how "hard" small organisms like bacteria may "try" to swim straight, they are prevented from doing so by the surrounding water molecules. After about a second of swimming in an approximately straight line—and the duration of each run is random—the flagella separate and start rotating independently in a clockwise direction. This causes the bacterium to move in an erratic fashion, appropriately called the tumble. Like the run the duration of the tumble is random, but lasts on average about one-tenth of a second, and then another run is initiated. Remarkably, such an infinitesimally small organism can sample its chemical environment as it swims forward. If it detects an increase in the concentration of useful molecules, from one temporal sample to the next, it prolongs the duration of its run. But if the concentration of favorable molecules decreases, or if it detects increasing concentrations of undesired molecules, the run is shortened. In this way the bacterium is able to move toward

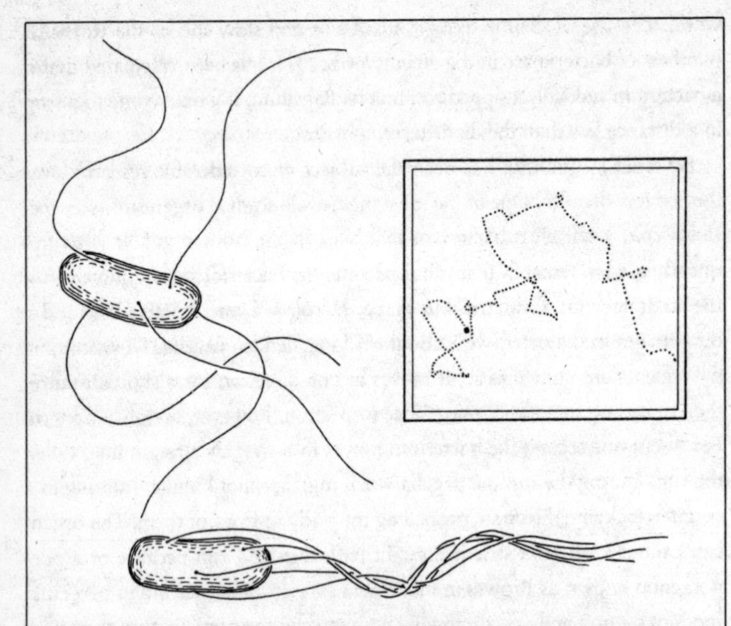

The bacterium Escherichia coli. *The flagella, which number up to six, rotate independently during the tumble phase of locomotion, which is consequently erratic. During the run phase, however, the flagella work together as a bundle, propelling the bacterium through the water, body first. Locomotion is therefore characterized by fairly straight runs, interspersed with erratic tumbles, as shown in the inset.*

favorable chemicals in the water and avoid unfavorable ones. The high viscosity situation makes this the only mechanism the bacterium has for increasing the concentration of desirable molecules in its vicinity. Wafting molecules toward it, in the way that a copepod attracts food particles with a feeding current, is not an option because inertia cannot be imparted to the water. Their only other alternative is to be non-motile, like many other bacteria, and to wait for the desired molecules to move past them by diffusion.

The rotary motor that drives the bacterial flagellum has a diameter of about 25 nanometers (one-thousandth of a micrometer) and generates a miniscule amount of power. On a weight-for-weight basis, however, the power is staggeringly high, and has been estimated at 10 horsepower per pound (about 3,500 watts per kilogram). To put this into perspective, the maximum power output of a pigeon's pectoralis muscle is about 120 watts per kilogram, some thirty times less.

When a plankton sample is collected close to land sometime between spring and fall, it contains all manner of larvae of littoral and benthic species. Most of these larvae are fairly large, up to about 2 millimeters, and motile. Because of their large size the locomotory organs are cilia rather than flagella, which are the prerogative of the smallest organisms. They include the veliger larvae of molluscs, that look something like butterflies, the barrel-shaped trochophore larvae of bristleworms (polychaetes), and the many kinds of pluteus larvae of sea urchins, starfish, brittle-stars, and sea-cucumbers (collec-

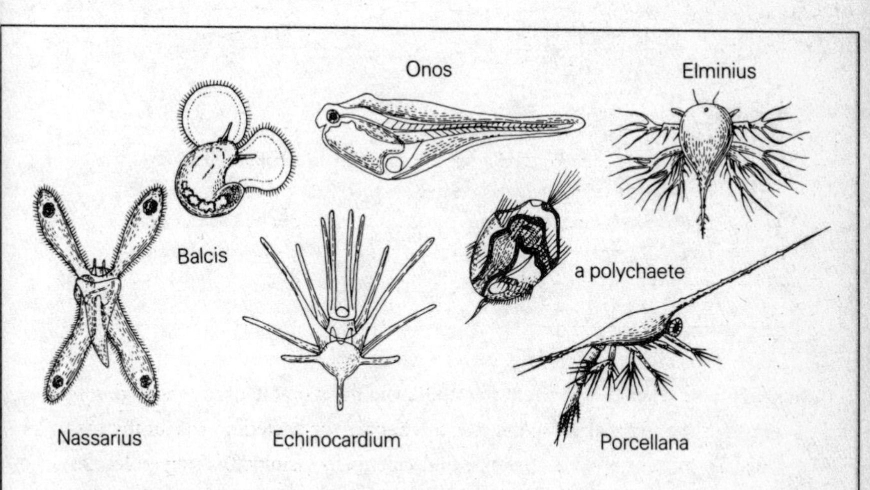

A diverse array of animals have planktonic larvae. From left to right: *a late veliger larva of the whelk* Nassarius, *an early veliger larva of the snail* Balcis, *a pluteus larva of the sea urchin* Echinocardium, *the larva of a fish,* Onos, *a trochophore larva of a polychaete (bristleworm), a zoea larva of the crab* Porcellana, *a nauplius larva of the barnacle* Elminius.

tively called echinoderms). This gypsy life, drifting with the tide, lasts only days, possibly weeks. Then the lucky ones, those that were not snapped up for food, settle down on terra firma, never to go wandering again. For the larval fishes that survived their sojourn in the plankton, though, the adventure has only just begun. Turning their backs on an orderly world dominated by viscous forces, they head off to the unpredictable realm of high Reynolds numbers, where inertial forces reign supreme.

LIFE IN THE FAST LANE

ONE MAGICAL NIGHT in Galápagos, moored beneath the stars, I stood aboard a motor yacht and watched a sea lion chase a flying fish. What transformed wonder into fantasy was the spectacular bioluminescence, the likes of which I had never seen before, or since. Each time the fish or sea lion broke surface, the sea exploded in phosphorescent fire. Jewels of light dripped from scales and fur, and an ephemeral blue trail was blazed in the sea behind them. To separate reality from imagery, the flying fish was the perfect icon of the differences between moving at high Reynolds numbers in water and in air. Water is more than eight hundred times denser than air and, since drag at high Reynolds numbers is directly proportional to density, the sudden reduction in drag when the flying fish breaks the surface enables it to catapult into the air. Spreading its pectoral fins, which function as wings, generates lift to prolong its flight. For good measure its modified tail gives a last propulsive flick against the water. The fish's sojourn in the air lasts only a few seconds, but the distance covered in that time is far greater than it would have been in water, lengthening its lead on the sea lion. Returning to the sea, with its eight-hundred-fold increase in drag, is like hitting a wall, and the fish takes a few moments to get under way again. Meanwhile the charging sea lion, whose own body has been breaching the surface like a porpoise, closes the

A flying fish in flight.

gap. Tail thrashing and body undulating, the fish builds up enough momentum to break clear of the water again, taking to the air on its finny wings in another explosion of light. And so the chase continues. The sea lion never does catch the flying fish.

As well as being over eight hundred times denser than air, water is also more viscous, by a factor of about 55 at 20°C. We have seen that the Reynolds number expresses the relative importance of inertial and viscous forces, and is the product of length, speed, and the ratio of density to viscosity. Since this ratio is about fifteen times higher for water than for air (800 ÷ 55), the Reynolds number of a body moving in water is fifteen times higher than it would be in air at the same speed. Aquatic animals therefore do not have to be especially fast nor large to achieve rather high Reynolds numbers—the animals we will be concerned with here have values ranging from 10^5 to over 10^7.

Since water is eight hundred times denser than air it follows that objects immersed in water are correspondingly more buoyant than they are in air. The upthrust on an object in water is equal to the weight of the fluid displaced, a fact that came to Archimedes, so the apocryphal tale goes, while he was soaking in the tub. The revelation caused him so much joy that he ran through the streets shouting Eureka! The buoyancy of air is negligible, except for lighter-than-air objects like helium balloons, but is considerable for water. Most animals have a density close to that of water, so aquatic animals are weightless when they are in water and can grow to enormous sizes. It is no coincidence that the largest living animal, the blue whale (*Balaenoptera musculus*), is aquatic. Reaching lengths in excess of 100 feet (31 meters) and weights over 200 tons, it is the largest animal that has ever lived, bigger even than the largest sauropod dinosaurs.

Fishes were the earliest vertebrates. Over the course of hundreds of millions of years they have evolved a diverse and dazzling array of adaptations to the aquatic environment. They are also the largest group of living vertebrates, with more species than any other, and have exploited every conceivable marine and freshwater niche. Without going into the nuances of modern classification, there are two distinct types of fishes, cartilaginous and bony, and they have had long and separate evolutionary histories. Nevertheless, many of them have evolved identical solutions to the same problems, thereby endowing them with similar features. Certain sharks, for example, have evolved a stiff *lunate* (crescent-shaped) tail with a high aspect-ratio, just like that of the swordfish and its relatives. This phenomenon of similar forms for similar functions is known as *convergence*. The cetaceans, relative newcomers, first appeared in the fossil record during the Eocene, about 50 million years ago. They, too, have evolved a similar tail, and are therefore said to be convergent on certain sharks and bony fishes for their lunate tails.

Sharks are among the most graceful of all swimmers and are a good starting point for our survey of high Reynolds number swimming. Like most other people, I have a long-standing fascination with them, but until recently my closest encounters with sharks had been in the laboratory and in restaurants. I finally got to observe them in their own domain while fishing for swordfishes aboard a research vessel off the Canadian coast. We did not catch many swordfishes, but we did land a lot of sharks, the first one being a blue shark (*Prionace glauca*). These magnificent animals, belonging to a group with the evocative name of requiem sharks, are named for their color, which changes from a steely blue in the dorsal area, to shades of silver blue, to a white underbelly. Their bodies are so beautifully streamlined it is easy to imagine them slipping through the water as smoothly as silk without making a single ripple. This, of course, is the purpose of the streamlined shape, the idea being to maintain a laminar flow in the boundary layer and beyond. Under such ideal conditions the drag on the shark would be minimal, and entirely attributable to the viscosity of the water in the boundary layer. This ideal is seldom realized in the real world, however, and part of the boundary layer—often a large part—is always turbulent.

The blue shark has a remarkably supple body, and its long asymmetrical tail feels as pliant as rubber. Such flexibility lends grace to its movements and imparts a wonderful fluidity to its swimming, enhanced by remarkably long and slender pectoral fins. Sharks have a *heterocercal* tail, meaning that the larger upper lobe is supported by an upturned segment of the vertebral column. According to conventional wisdom this configuration implies that the upper lobe is stiffer than the lower lobe, and consequently trails behind it when the tail sweeps from side to side. The flexible lower lobe therefore functions as an inclined plate, giving the heterocercal tail its characteristic upthrust when driving the shark forward through the water. This upthrust acts behind the center of gravity (center of mass). It is balanced by an upthrust in front of the center of gravity generated by the paired pectoral fins, which act as inclined plates as they cut through the water. In this way the shark, whose body is usually denser than seawater, is prevented from sinking. This standard explanation of shark locomotion has been textbook dogma for half a century, but investigations during the late 1970s exposed it as a faulty over-

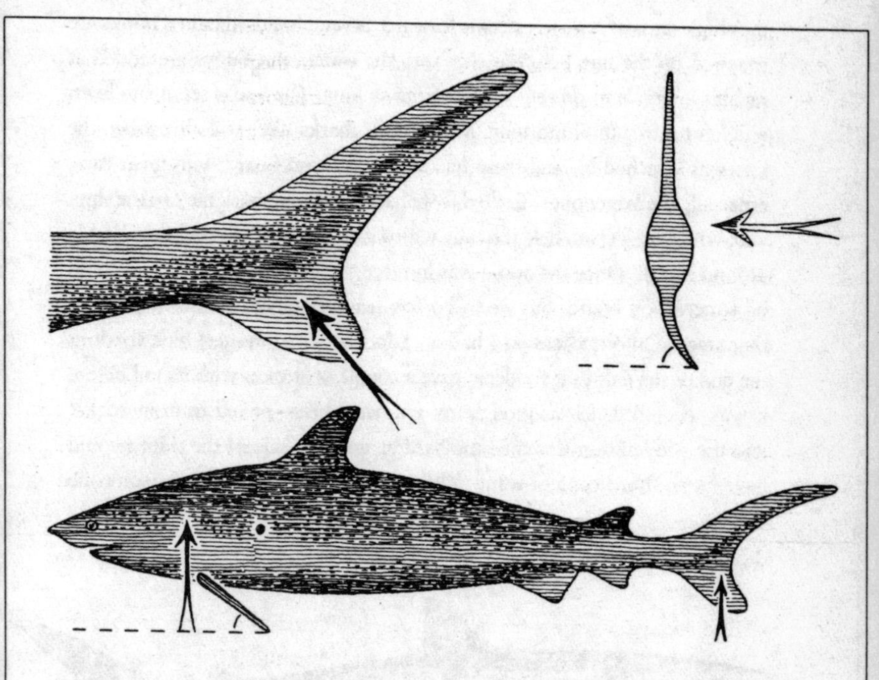

The lower lobe of a shark's tail is often more flexible than the upper lobe. Consequently, when it sweeps from side to side, it is deflected by the water pressure. It therefore functions as an inclined plate, shown here in posterior view (top right)*. This results in an upthrust on the tail* (top left) *as well as a forward thrust. This upthrust, acting behind the center of gravity, is balanced by an upthrust in front of the center of gravity generated by the pectoral fins set at a positive angle of attack. This standard explanation of swimming in sharks is an imperfect oversimplification.*

simplification of a complex mechanism. I became interested in the issue because ichthyosaurs, a group of extinct marine reptiles that I study, have a reversed heterocercal tail—the vertebral column tilts downward into the lower lobe.

Our catches during this fishing trip were all taken on a long-line, a commercial method of fishing where several kilometers of line are paid out from the ship's stern as it slowly steams forward. Several hundred baited hooks are snapped on the line before it slips into the water, and buoys are added at regular intervals to prevent the line from sinking. The line is set at sundown and left to drift until morning. Most of our sharks were still alive when the line was winched in, and these had to be taken on board with great care, especially the larger ones. To do this we had to pull the shark next to the ship, and, while one person held it steady with a gaff, a second slipped a steel cable around its tail. Once the noose was in place the shark was secure and could be winched on board, but we had a few tense moments before we reached that stage. On one occasion I had a 12-foot-long (3.7-meter) blue shark at the end of the gaff that suddenly gave a couple of strokes with its tail before it was secured. It felt as though my arm were being pulled from its socket and the acceleration slammed me hard against the side of the ship. I could have let go, but I did not want to lose the shark. A moment later someone

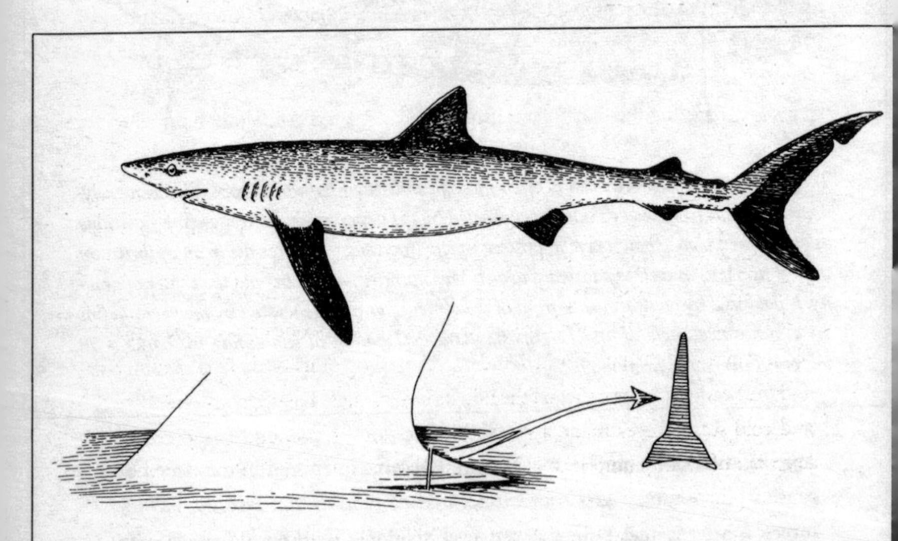

The blue shark, Prionace glauca. *The closeup shows details of the fairing on the trailing edge of the dorsal fins, including a transverse section (*bottom right*).*

slipped the noose over the tail and immobilized it. As the shark was winched aboard it arched its powerful body, first to one side and then to the other, completing a loop by touching its tail with its snout. This remarkable display of brute strength and flexibility was quite poignant. A big part of me wanted the magnificent creature to be back in the open sea where it belonged.

Like many other sharks, the blue shark has two dorsal fins and one anal fin. These too, have streamlined profiles, like the paired pectoral and pelvic fins. Closer inspection of the dorsal and the anal fins revealed something that I had not noticed in the preserved specimens back at the Museum. The trailing edge of each fin forms a loose flap. Although this is not permanently attached to the body, it is expanded into a fairing that is contoured to the shape of the body, with which it maintains contact. The fairing is lubricated with mucous that appears to maintain its close contact with the body. As I knelt on the deck marveling at this fine piece of engineering, mindful that the shark could still bite, the trailing edge flap began to move, first to one side then to the other, like a ship's rudder. As the trailing edge flap moved, the fairing tracked across the body with electronic precision. The movements of the trailing edge flap probably have a stabilizing role or help to guide the shark as it swims.

The next shark we winched aboard was something completely different. This was a mako shark (*Isurus oxyrhynchus*), a member of the same family as the great white shark (*Carcharodon carcharias*). It thrashed about savagely at the end of the noose, snapping its jaws at everything in sight with unbridled malevolence. In marked contrast to the blue shark, its body was quite stiff. No matter how hard it thrashed, the lateral arching of its body was minimal and the head never approached the tail. Its sharp, jagged teeth, pointed snout, and cold staring eyes made a marked impression on me, and I resolved to keep out of its way until it was decidedly dead.

When fishes are in the water they irrigate their gills with freshly oxygenated water by pumping with their mouth and throat, as can be seen by observing fishes in an aquarium. Like other vertebrates, fishes are able to exercise aerobically, supplying oxygen to their muscles at a sufficiently high rate to sustain their activity. Experiments performed in flumes—the aquatic equivalent of a wind-tunnel—have shown that as fishes swim faster they ventilate their gills more rapidly, just as we breathe more rapidly when running. But large fishes soon reach a point where they switch over to ram-ventilation, opening their mouths and letting the water rush over their gills. However, the switch-over may have more to do with hydrodynamics than respiration. Sharks and teleosts (advanced bony fishes to which most modern species belong) appear to switch over to ram-ventilation at about the same swimming speeds (0.6 to 1.3 miles per hour). But some evidence indicates that teleosts may be able to maintain higher aerobic swimming speeds than sharks.

Data for the running speeds of terrestrial animals are meager enough, but the situation for aquatic animals is far worse. The reason for this is obvious—we spend most of our time above the surface; aquatic animals spend most of their time beneath it. When our paths cross, though, we have an opportunity to measure their speeds. Dolphins often swim beside ships, as do whales, and fishermen who troll for sport fish like marlin can see how fast their boat is traveling when a marlin catches the bait. To our shame, we have tainted data on how fast harpooned whales attempt to swim away from whaling ships. Collectively, these data do not amount to much, and some have obvious problems with accuracy. For example, dolphins swimming beside ships often benefit from the water displaced by the hull, and their speed is hence exaggerated. A particularly exaggerated speed happens when dolphins ride the bow wave, sliding down the crest of water that the ship drives in front of its bow. And what is the value of knowing how fast a terrified whale can swim with a large part of its body blown away? Knowledge that a sporting marlin can chase after a fisherman at 40 miles per hour is of limited value too because we do not know whether such bursts of speed are part of the fish's normal behavior, or whether the individual marlin is typical of its species.

What would be most informative is a fish's maximum sustainable speed. Since it is rather difficult to establish this in the field, numerous determinations have been made in the laboratory by using a flume. For convenience these assessments have usually been of species that are easy to obtain and maintain, like salmon, trout, carp, and goldfish. Because of size constraints imposed by the flume, they are all fairly small. For all these reasons, then, our knowledge of swimming speeds in fishes and other aquatic animals is limited.

When the mako shark finally succumbed I took a closer look. As in the blue shark, the dorsal and anal fins had trailing edge flaps with fairings contoured to the body. And what a stiff body it was compared with the other shark! The tail was as stiff as a plank of wood, with no difference between the upper lobe supported by the vertebral column and the unsupported lower lobe. It was also more nearly symmetrical; the upper and lower lobes were of similar shape and size, like a crescent moon. The caudal peduncle, the narrow "wrist" region that lies in front of the tail, was much stiffer than in the blue shark and bore a pair of distinct and rigid lateral keels. These may stabilize the fish by correcting for pitching. Alternatively the keels may provide lateral stiffness, or reduce the drag on the caudal peduncle as the tail beats from side to side. As in many other sharks, the mako has a prominent pit on the dorsal aspect of the peduncle, immediately in front of the tail, together with a transverse crease. These suggest that the tail may have a limited amount of dorsoventral (up and down) movement. I have long suspected that sharks may be able to exercise a small degree of dorso-ventral control on the pitch of their tail, which would enable them to modify the vertical forces generated. If this were so, the tail could adjust itself to generate upthrust, downthrust, or solely forward thrust. I hasten to point out that my suspicion is entirely speculative, but an investigation would make an interesting research project. The mako and its relatives, all characterized by stiff lunate tails and bodies, are collectively named lamnid sharks, borrowing from the generic name of one of their members, the porbeagle (*Lamna nasus*, from the Greek, *lamna*, a fish of prey), which we also caught that day.

The function of the lamnid shark's stiff body is apparent when its swimming movements are compared with those of other sharks. Lamnid sharks are sparing in their body movements. Most of the forward thrust of swimming comes from their tail, which functions as an inclined plate. A blue shark, in contrast, throws its entire body into lateral undulation when it

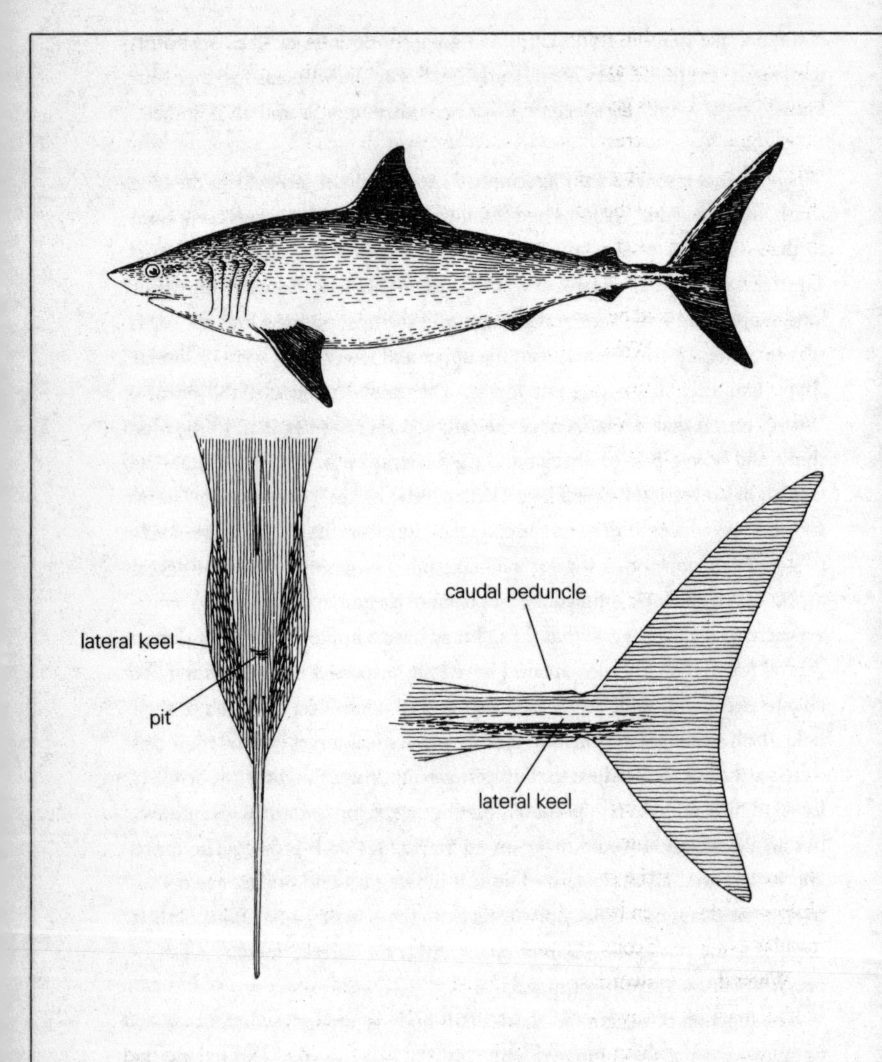

The mako shark, Isurus oxyrhynchus, *showing details of its lateral keel and caudal peduncle.*

swims. The wave that moves from head to tail down its body reacts against the water's resistance and propels it forward, adding to the tail's thrust. Elongated fishes like eels generate thrust from body undulations, whereas others, like lamnid sharks, create most of their thrust with the tail. Keeping the body stiff and concentrating movement in the tail greatly reduces drag. Flexible-bodied sharks enjoy the advantage of pushing against more water and therefore accelerating faster—but the additional drag costs are high.

When an animal is actively swimming the drag on its body is several times higher than it would be for a rigid model of similar shape and size. However, as zoologists Paul Webb of the University of Michigan and Robert Blake of the University of British Columbia have shown in their many important contributions to our knowledge of swimming, an animal can achieve a similarly low drag by temporarily ceasing to swim and coasting instead. The alternation of active swimming and passive coasting is characteristic of many swimmers, and can lead to energy savings of up to about 80 percent. Low Reynolds number swimmers, we have seen, are unable to coast and are therefore unable to exploit any energy saving in this way.

My research over the years has centered on ichthyosaurs, a remarkably successful group of reptiles that flourished in the sea for most of the Mesozoic Era, from about 250 million to 100 million years ago. Several thousand specimens are known, many in excellent condition; some even have the body outlines preserved as a carbonaceous film. Hence we have a good idea of what they looked like in life. Their closest living analogue are the scombroid fishes, as exemplified by the swordfish (*Xiphias gladius*). One ichthyosaur, named *Eurhinosaurus*, even has a shortened lower jaw just like a swordfish, a spectacular example of convergence.

When the first swordfish was hoisted aboard I was reminded of a dream I had many years ago, on the eve of my Ph.D. oral examination. I had dreamed that a fishing boat operating off the Cornish coast of England had caught a fresh ichthyosaur. In my dream all I wanted to do was to finish my oral as quickly as possible, before somebody examined the specimen and discovered that everything I had inferred about their biology was wrong!

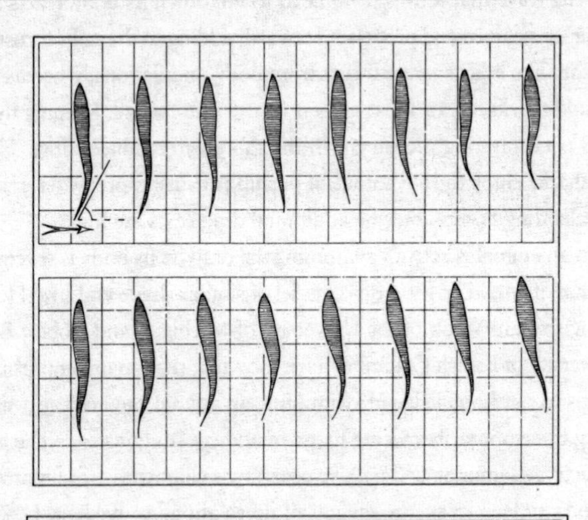

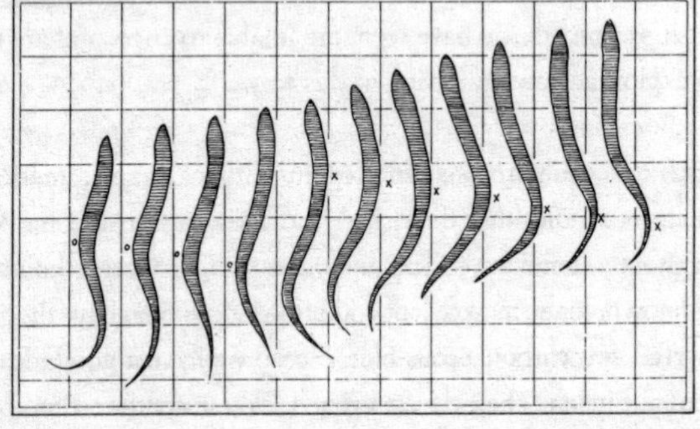

Time elapse sequences of fishes swimming. Top and middle: *In the whiting* (Gadus merlangus), *as in many other fishes, forward thrust is obtained both from the tail functioning as an inclined plate as it moves from side to side, and from the undulation of the body.* Bottom: *In long-bodied fishes like the eel* (Anguilla vulgaris) *the entire thrust of swimming is obtained from body undulation. The circles and crosses mark the regions of maximum curvature. The waves of undulation move down the body, acting against the resistance of the water to propel the fish forward.*

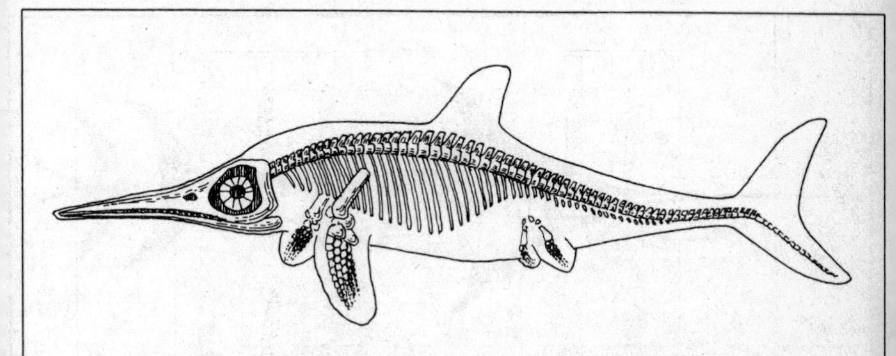

An ichthyosaur, an extinct marine reptile that lived during the Mesozoic Era. This example is one of the well-preserved specimens from the Lower Jurassic of Holzmaden, in southern Germany. Some of these specimens have the body outline preserved as a carbonaceous film so we can be fairly sure of their body shape.

Swordfishes not only look very much like ichthyosaurs, they are also big. The specimen lying on the deck was 10 feet long (3.2 meters) and weighed 442 pounds (201 kilograms), but they grow much larger—in excess of 14 feet (4.3 meters). One striking feature, not apparent in illustrations which invariably show lateral views, is how wide they are; indeed, they have an almost square cross section. The pectoral fins have a high aspect ratio and project stiffly downward from the sides of the body at an oblique angle. There are no pelvic fins and the first dorsal and first anal fins are stiff, with high aspect ratios like the pectorals. Just in front of the crescentic tail, which is strikingly long, narrow, and stiff like the other fins, are a small second dorsal and an equally small second anal fin. Close inspection revealed that these two small fins had a trailing edge flap and contoured fairing as in the sharks. What is more, the caudal peduncle has a lateral keel, like that of the mako shark, providing yet another example of convergence. The swordfish's sword is dorso-ventrally flattened, with edges so sharp that you can cut yourself on it if you are not careful. Precious few observations have been made on living swordfishes, but they have been seen swimming through schools of fishes,

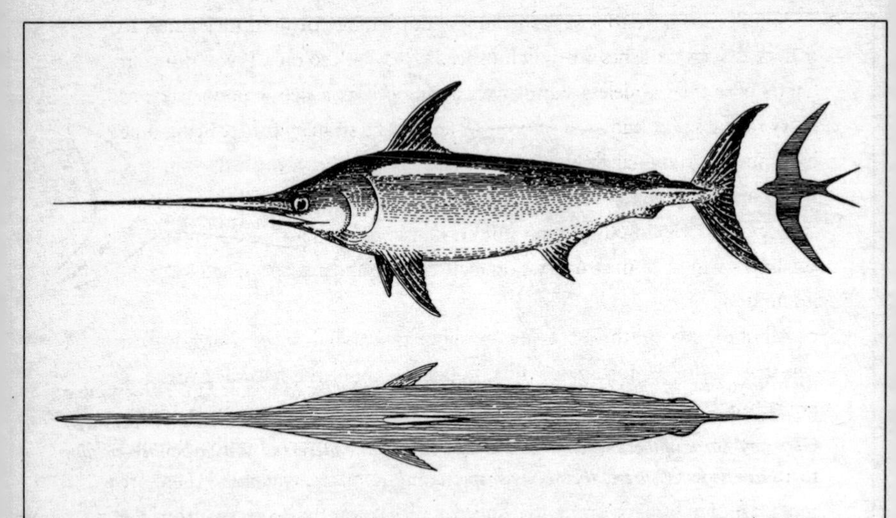

The swordfish, Xiphia gladius. *Notice the lateral keel, high aspect-ratio fins, and stiff crescentic tail, which is similar in shape to the wings of the swallow* (top right). *Viewed from above* (bottom)*, the body has a remarkably rectangular profile.*

immobilizing them with slashes from their swords. This is supported by evidence from stomach contents where fishes with slash wounds have been found. It has also been suggested that the sword has a hydrodynamic function, reducing drag.

Swordfishes, like most other scombroids (including mackerels), rely on ram-ventilation to provide a sufficient water flow over their gills, and they have to keep swimming all the time. The large swordfish we caught might have been hooked many hours previously and was therefore dead, but even if it had been alive I doubt it would have thrashed its body very much because the vertebral column is so stiff. This is in marked contrast to its close relative the sailfish, which can throw its remarkably flexible body into all manner of contortions when it is hooked on a fishing line. The swordfish body was not only stiff but hard, so much so that we used it as a temporary bench—like sitting on a log—while we worked on another specimen.

You do not have to know anything about biology or fluid mechanics to deduce that swordfishes are built for speed. They look so racy. If you run your hands over their scaleless skin it feels as smooth as a neoprene wetsuit; as every racing sailor knows, a smooth surface is one strategy for reducing drag by promoting laminar flow in the boundary layer. The swordfish's thin fins, with their sharp edges, are reminiscent of jet fighter wings, and this is also a strategy for reducing drag. The stiff crescentic tail has a shape similar to a swallow's wings; both structures oscillate rapidly and are optimized for minimum drag.

All of the swordfish's attributes are suggestive of high speed, but how fast can they swim? Unfortunately little is known about the natural history of swordfishes, including their swimming speeds, because they are so rarely seen. Large females are often sighted during the summer months sunning themselves on the surface, but few swimming data are available. There are more data for other scombroids, and the impression that emerges from the literature is that they can swim as fast as 80 miles per hour (130 kilometers per hour). But the highest recorded speeds are about 47 miles per hour (76 kilometers per hour), which is still remarkably fast. These are burst speeds, and just because an animal attains such high speeds does not necessarily mean that it does so very often, nor that it can keep it up for more than a few seconds. Lions can run fast, up to about 37 miles per hour (60 kilometers per hour), but they spend most of their time padding about slowly, or lying down resting. So how fast *do* these speedy-looking scombroids swim on a day-to-day basis?

Using an electronic speed-recording device with a radio transmitter, Barbara Block, a researcher at Stanford University, and her colleagues measured the swimming speeds and diving depths for three blue marlins (*Makaira nigricans*) off the coast of Hawaii over periods ranging from 25 to 120 hours. Remarkably, the marlins spent more than 97 percent of their time cruising at speeds of less than 2.7 miles per hour (4.4 kilometers per hour); their top speed was only 5 miles per hour (8 kilometers per hour). Pelagic (open sea) sharks do about the same speed. Blue sharks, between 7 and 9 feet (2.2 to 2.8 meters) long, commonly swim at 0.9 to 1.6 miles per hour (1.5 to 2.6

kilometers per hour), with top speeds of 4.5 miles per hour (7.2 kilometers per hour) for short bursts. A 6-foot (1.8-meter) mako shark swam at an average speed of 2 miles per hour (3.2 kilometers per hour) over a twenty-four hour period and had a maximum speed of only 3.4 miles per hour (5.5 kilometers per hour). In a tagging investigation using radio transmitters, five swordfishes were monitored for five days. By charting their movements the investigators could estimate their swimming speeds, which did not exceed 2.2 miles per hour (3.6 kilometers per hour). The same techniques give estimates for sustained swimming speeds for the bluefin tuna (*Thunnus thynnus*) of 3.3 miles per hour (5.4 kilometers per hour), which accords with flume results obtained by Jeffrey Graham and Heidi Dewar of the Scripps Institute of Oceanography.

What should we make of these seemingly fast pelagic fishes that appear to spend most of their time cruising along at a leisurely pace of only 2 or 3 miles per hour? Drag forces are high in water because of its high density, and these increase with the square of the speed at high Reynolds numbers. So if a swordfish accelerated from 2 to 8 miles per hour, the drag force would jump by 16 times. Here is reason enough for putting a limit on cruising speeds. But another aspect of drag has to be taken into account, namely the phenomenon of critical Reynolds numbers. Briefly stated, as Reynolds numbers rise a critical point is reached where the flow in the boundary layer changes from being laminar to turbulent, causing a dramatic rise in drag. The critical Reynolds number at which this transition occurs depends on the body's shape and is probably in the order of 10^6 for the fishes we have been considering. Keeping speeds low maintains low Reynolds numbers and it has been suggested that the reason why large pelagic fishes swim at low cruising speeds is to avoid reaching their critical Reynolds numbers, thereby avoiding the increased drag accompanying a turbulent boundary layer. Large pelagic fishes spend much of their time swimming in the cold deeper layers of the ocean. Clement Wardle, a British biologist at the Scottish Office of the Agriculture and Fisheries Department, has suggested that this behavior keeps Reynolds numbers below critical values. This is because water viscosity rises with lower temperatures and Reynolds numbers are inversely proportional to viscosity.

Another factor that may limit cruising speeds is respiration. For sustained swimming a fish must supply sufficient oxygen to its muscles for them to contract aerobically. This requires not only an adequate flow of water over the gills, but also a matching performance from the cardiovascular system and from the muscles themselves. All animals have an upper limit to their aerobic performance (mine is a modest jog), and this factor alone may account for the ceiling on cruising speeds. This idea is supported by flume experiments that show rather low sustained swimming speeds. For example, the maximum value recorded for twenty leopard sharks from 14 inches to 4 feet (0.35 to 1.21 meters) long, was a modest 2.2 miles per hour (3.5 kilometers per hour). Compare this with 2.2 miles per hour (3.6 kilometers per hour) for a 31-inch (0.8-meter) cod and 3.1 miles per hour (5 kilometers per hour) for a 21-inch (0.54-meter) salmon.* Because maximum sustained swimming speeds appear to vary with body size, and because few data are presently available, the respiratory explanation is only tentative.

While cruising speeds of the order of 2 or 3 miles an hour may seem disappointingly slow for large pelagic animals, they can still cover enormous distances if given sufficient time. A black marlin tagged off the coast of southern California, for example, was recaptured three months later off Peru, a distance of over 3000 miles (4800 kilometers). If the fish swam continuously, this distance could have been covered at an average speed of under 2 miles per hour (3 kilometers per hour).

A visit to a good fish market can be an informative as well as a gastronomically rewarding experience. The best one I have ever visited was in Barbados, a gigantic market filled with wonderful people and exotic fishes. What interested me most were the large pelagic fishes—tunas, marlins, and sharks—most of which had been cut transversely to reveal the distribution of red and white muscle fibers. In some fishes, like marlins and sharks, the deep red muscle was restricted to a few small blocks lying just beneath the skin, contrasting sharply with the rest of the flesh, which was white. Tunas, in contrast, had blocks of red muscle lying close to the vertebral column; the rest of the flesh had a slightly pink tinge. In all instances the red muscle com-

*Data from Graham et al. 1990, and Wardle 1977.

prised a small fraction of the whole, so if you did not like red meat there would be hardly anything to trim off your steak.

Experiments in flumes have shown that during sustained (aerobic) swimming sharks use only their red muscles. Bony fishes are similar, but some recruit white fibers too. Since large pelagic fishes appear to spend most of their time swimming aerobically, their white fibers, the bulk of their muscle mass, are rarely utilized. It has been suggested that fishes can afford the luxury of carrying so much excess baggage only because of the water's buoyancy. Terrestrial animals could not afford to do so. However, the fact remains that on occasion these fishes need to sprint, either to catch their prey or to avoid becoming prey themselves. So it is a specious argument to suppose that the white muscle fibers are in excess of their needs.

I have reviewed some of the mechanisms used by high Reynolds number swimmers to minimize their drag: streamlined bodies, fins with streamlined profiles, high aspect-ratio tails and fins, and smooth skins. Other strategies exist; the best exemplars of these are the tunas. The tunas, the largest species of which reach lengths of 14 feet (4.3 meters), are paragons of engineering excellence. Their body's beautifully streamlined form is as smooth as that of the swordfish. The swordfish's fins project stiffly from the body, but the tuna can tuck most of its fins out of the way. The pectoral fins can be flattened against the body, and, although the fins are thin, the body is recessed for their reception so that they lie absolutely flush with the surface. The anterior dorsal fin can completely collapse into a longitudinal groove, and a pair of grooves receive the collapsed pelvic fins. The free margin of the operculum, the flap that covers the gill openings, fits into a shallow recess on the side of the body—yet another flush fit. Even more remarkable, the eyeball faithfully conforms to the body's contours. Run a finger along the body, from the tip of the snout to the root of the tail, and it feels smooth all the way. When swimming the tuna keeps its collapsible fins flush with the surface, thereby reducing drag, and extends them only when required, as when changing directions. Their pectoral fins, though, are important for generating lift and are kept extended except during rapid bursts of speed. I have watched their diminutive relative, the mackerel, swimming in a large marine tank. The mackerel incessantly flicked its fins up and down each time it changed direction.

As Reynolds numbers increase the boundary layer not only becomes turbulent but begins separating from the body's surface. As mentioned earlier, separation begins posteriorly and moves forward with growing Reynolds numbers. Separation can be delayed by stirring up the fluid flow, and making it turbulent. This is because, at high Reynolds numbers, a turbulent boundary layer adheres to the surface more closely than a laminar one. The result is a smaller wake and less pressure drag. The medium finlets in the tuna's caudal region may function as vortex generators, stirring up the boundary layer in the posterior region where separation is most likely to begin. The tail regions of commercial airplanes are prone to airflow separation, a problem resolved by the use of vortex generators. These devices, seen in the tail regions of several aircraft including the Boeing 727 and 737, take the form of small rectangular tabs set obliquely to the direction of flow. The obvious disadvantage of vortex generators, whether on the tails of aircraft or tunas, is that at lower Reynolds numbers they increase drag by causing an otherwise predominantly laminar situation to become turbulent. But the disadvantage is offset by the reduction in drag at high Reynolds numbers. An alternative explanation for the tuna's finlets is that they function to modify the flow of water passing over the rapidly beating tail.

Fishes are slippery to the touch because they produce mucous that is discharged onto the skin. Mucous has several functions, including acting as a barrier to water (freshwater fishes absorb water because the osmotic pressure of their body fluids is higher than that of their environment, whereas marine species lose water for the opposite reason) and protection against parasites and abrasions. Another function of mucous, used by many species but apparently not by scombroids, is drag reduction. The active agent is a water-soluble polymer; there are many of these, and they are functional even at low concentrations. One of the most effective of these polymers is produced by the Pacific barracuda, a fast predator; experiments have shown that a 5 percent solution in seawater reduces drag by 65 percent. A United States naval research establishment conducted research on fish mucous, doubtlessly to im-

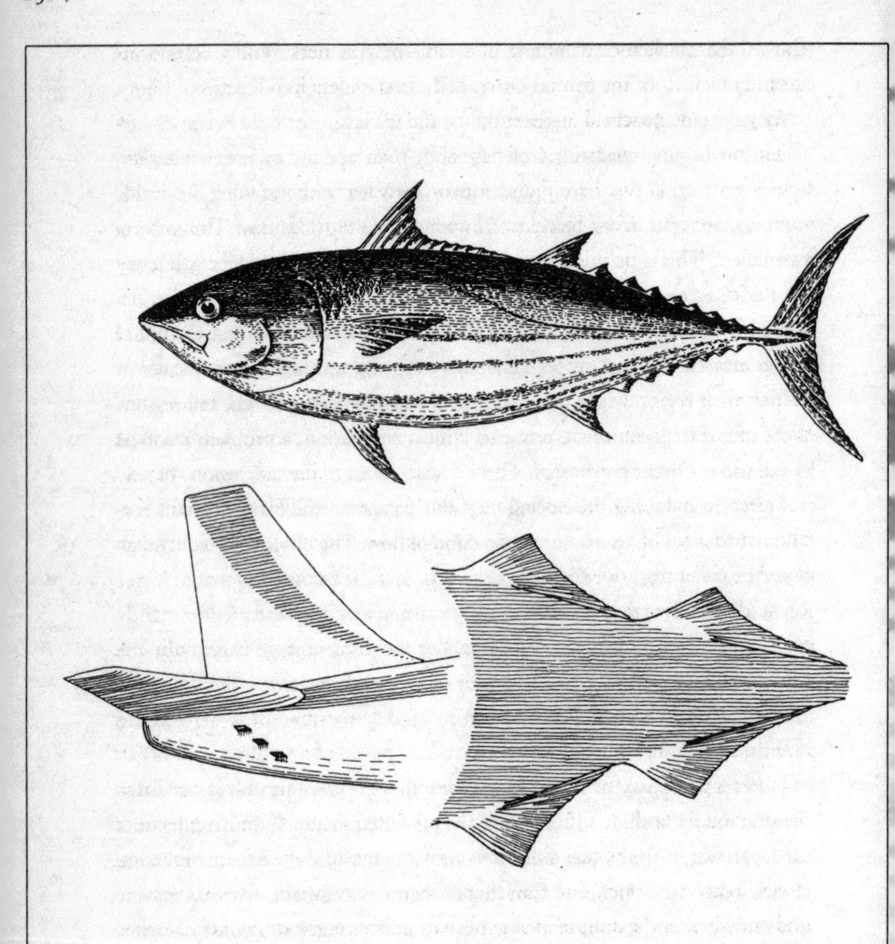

The largest tuna species reaches lengths of 14 feet (4.3 meters). Notice the beautifully streamlined profile, the stiff crescentic tail, and the high aspect-ratio fins. A series of finlets running along the middle of the back and rump (detailed at bottom right) *might function as vortex generators to delay separation of the boundary layer. Vortex generators are used for this purpose in aircraft, as at the root of the tail of the Boeing 737* (bottom left). *Alternatively the finlets may function to modify the flow of water over the rapidly beating tail.*

prove the performance of submarines. But the polymers rapidly deteriorate after removal from the fish and lose their effectiveness in a matter of hours, so they have no practical application for the Navy.

The swimming mechanism of fishes like tuna and swordfish, whose stiff bodies generate thrust by rapidly oscillating their crescentic tails, is referred to as *thunniform* (from the Latin *thunnus*, for the tuna fish). Thunniform swimmers, which include the cetaceans, are optimized for cruising but not for rapid acceleration. Their shortcomings in acceleration can be demonstrated for some species of scombroids by catching and releasing them. On being released the fish wag their tails furiously without making any headway at first; then their tails begin to "bite" and they take off at speed. I have tried this experiment with mackerels. When held in the hand their tails oscillate rapidly, like a battery-operated toy, but when released into the water they do not seem to have any problems in making a fast start. This is probably because their propulsive thrust is not confined to the caudal fin but involves undulations of the body too.

In 1936 Sir James Gray, a leading light in the field of animal locomotion, published a paper on the swimming performance of a porpoise that posed a problem so baffling that it became known as "Gray's paradox." The kernel of the problem was that a porpoise achieved a speed of 23 miles per hour (37 kilometers per hour) for 7 seconds, a feat seemingly possible only if it reduced the drag on its body to a fraction of the predicted value. With the advent of the Cold War in the 1950s, interest in Gray's paradox shifted from biological circles to military ones, and the United States Navy began investigations to find out whether the drag-reducing mechanisms of cetaceans could be applied to submarines. The first step was to gather accurate data on the swimming performance of dolphins and porpoises, and this began in 1960 at the Naval Ordnance Test Station in California. Conventional submarines of that era operated at Reynolds numbers as high as 10^9, under which conditions the flow is probably turbulent throughout the entire boundary layer. Cetaceans, in contrast, were thought to be capable of maintaining laminar flow throughout much of the boundary layer. If such a degree of laminar flow could be achieved for submarines, drag would be drastically reduced and speed would be correspondingly increased for the same engine power. A clue to the drag-reducing mechanism of cetaceans lay in their skin, whose spongy texture possibly damped disturbances in the boundary layer before it became turbulent. Experiments were conducted with various resilient coatings, but no reductions in drag were achieved.

As more data were gathered it began to dawn on the researchers that cetaceans did not possess any special drag-reducing mechanisms at all. Indeed, their transient bursts of high speeds could be achieved with the boundary layer completely turbulent, as in submarines. Their only "secret" was that their bursts of power were more than three times higher than Gray had cal-

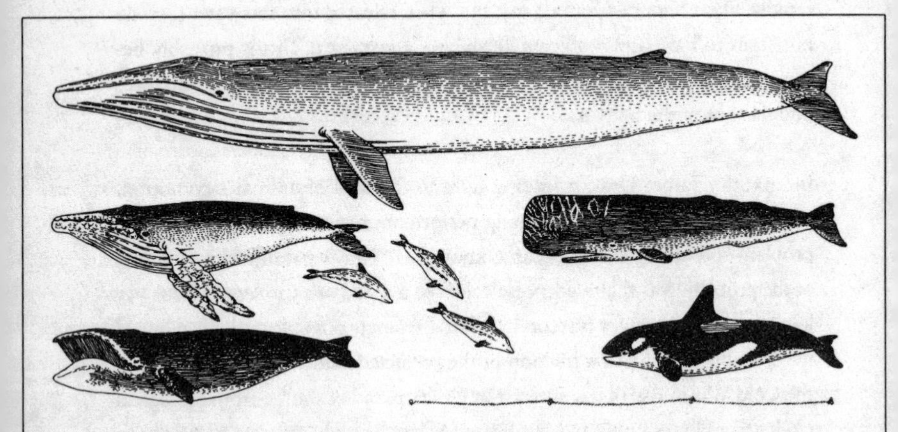

There are two major groups of cetaceans, the toothed whales (odontocetes) and baleen whales (mysticetes). The blue whale, the largest living animal (top)*, and the humpback whale* (middle left) *are both rorquals, easily recognized by the deep grooves in the throat region. The Greenland right whale* (bottom left)*, also known as the bowhead whale, is one of three living right whales, recognized by their deep, bow-shaped upper jaws. Dolphins and porpoises* (middle) *are the smallest cetaceans and belong to the same group as the killer whale* (bottom right)*. The sperm whale* (middle right) *is the largest toothed whale. The rope scale is knotted at 16-foot (5-meter) intervals.*

culated. Gray's estimate for his porpoise's power output was based on what a human athlete could maintain for a 15- minute period, whereas the cetacean's burst lasted only 7 seconds. In hindsight we can see where Gray went wrong, but his was a reasonable premise at that time. The porpoise's power output was therefore grossly underestimated during its short burst; hence the drag forces on its body were underestimated too. Just think how much cetacean research was stimulated by this simple oversight, not to mention how much money the United States Navy spent on rubberized paint!

Dolphins and porpoises belong to one of two groups of cetaceans called the odontocetes or toothed whales. The second group is the mysticetes or baleen whales, sometimes also called whalebone whales—all toothless. In terms of numbers of species the odontocetes are the largest group, and they range in size from 4 to 10 feet (1.2 to 4.0 meters) for porpoises and dolphins, to 25 feet (8 meters) for killer whales, and 50 feet (15 meters) for sperm whales. The mysticetes, with less than two dozen species, include the blue whale, the largest of all animals. This magnificent creature reaches lengths in excess of 100 feet (31 meters) and weights of over 200 tons. Such enormous sizes are possible only in the aquatic environment, where the body mass is supported by the water's buoyancy. Large forces have to be generated to drive these behemoths through the water, but these forces are considerably smaller than they would be on land. Compare the effort required to push a truck and a ship. There is no way I can push a 2-ton truck along a flat road, but I have tried pushing a 40-ton fishing boat through water and it is surprisingly easy.

Because of how easy it is to capture and maintain them in captivity, most of our data for swimming performance in cetaceans pertain to the smallest ones—the dolphins and porpoises. Their intelligence and readiness to participate in experiments has provided ample data. They can achieve bursts of speed of about 25 miles per hour (40 kilometers per hour), and they can maintain speeds of about 22 miles per hour (35 kilometers per hour) for about 10 seconds. Their maximum sustained swimming speeds appear to be closer to 7 miles per hour (11 kilometers per hour), though they may not cruise at such high speeds in the wild.

In the days of sailing ships maritime convention required steam to give way

A rorqual whale. Middle left: *closeup of the head with lower jaw lowered to show baleen plates.* Middle right: *a transverse section through the head.* Bottom: *a partially dissected view of the upper jaw to show the arrangement of the baleen plates.*

to sail. The convention no longer applies because modern cargo liners and oil tankers are incapable of making sharp turns, or of stopping in short distances. This is because of their enormous momentums, and the same is also true for the larger cetaceans. A dolphin may be as nimble in the water as a lamb in the field, but a 60-foot fin whale is an elephant on ice in comparison. So it should come as no surprise that all large whales feed on small organisms that they can scoop up indiscriminately without having to maneuver and catch individuals. Almost all whales in this category are mysticetes; the exception is the sperm whale, which feeds on schools of squid. There are two kinds of mysticetes, right whales and rorquals, and both are filter-feeders, using baleen as the straining device. Baleen, a horny material formed of the protein keratin, takes the form of thin plates, fringed by bristles, that hang down like curtains from either side of the mouth. Baleen is both resilient and strong. Prior to the invention of plastics, it was used in the manufacture of a wide variety of household goods. The bristles, for example, were used for making brushes and brooms, while the plates were made into combs, shoehorns, and boning for ladies' corsets.

Right whales, with only three living species (black, Greenland, and pygmy right whales), have great bow-shaped upper jaws that support a row of remarkably long baleen plates, sometimes exceeding 10 feet (3 meters) in length. These fit inside deep cheek flaps, so when the mouth is fully open no gaps are left at the bottom of the baleen. The whales feed by swimming slowly through the water with their mouths agape, filtering plankton from the water as they go. This form of filter feeding has been described as continuous ram feeding. Every so often the whale's tongue scrapes plankton from the inside of the plates prior to swallowing.

Rorquals, in contrast, lack a bowed upper jaw and consequently have shorter baleen plates. Their most striking feature is a series of deep grooves in the throat region that extends all the way back to the midriff, delimiting a distensible throat pouch. Instead of cruising along with their mouths open, rorquals feed by swimming up to a suitable patch of food items, opening their mouths, and scooping up the whole lot, water and all. The drag forces generated when that great barn door swings open must be enormous, and it is

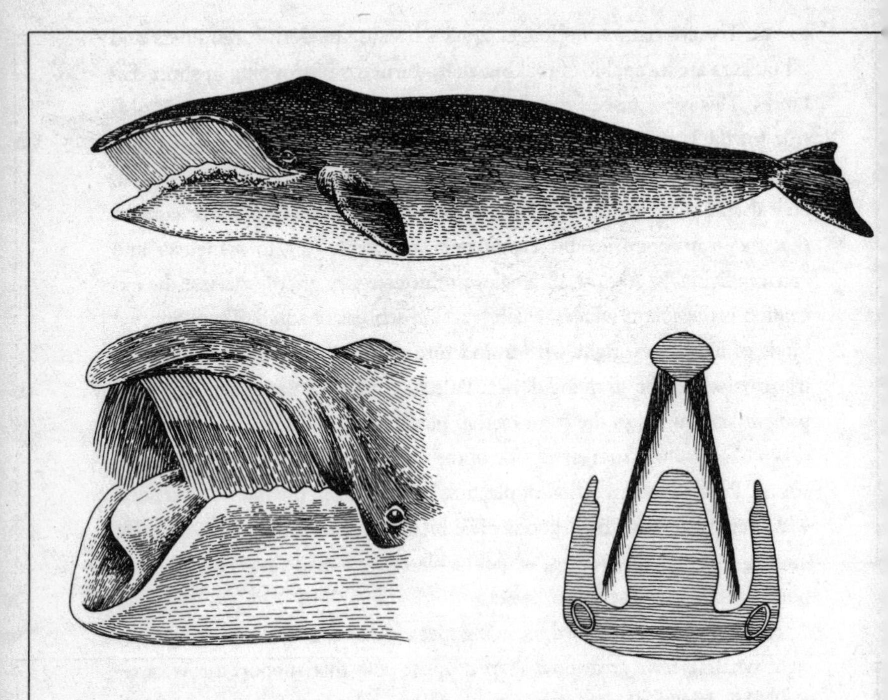

*A right whale showing its deeply arched upper jaws and long baleen plates (*bottom left*). The baleen plates slot into deep cheek flaps, as shown in the transverse section through the head (*bottom right*). Notice the huge tongue.*

largely the inertia of their gigantic bodies that keeps them moving forward. The drag forces distend the throat pouch so much that blue whales can engulf as much as 70 tons of water in a single scoop. The mouth is then closed and the water squeezed out at the sides, straining off the ingested food items. These intermittent ram feeders are quite catholic in their tastes, swallowing up fishes as well as plankton. They seek out schools of herring or mackerel, swallowing up hundreds of pounds of fishes in a single gulp. And as the door swings closed and day becomes night, the unsuspecting copepods continue their diatomic feast. But the herring world around them collapses into disorder as worlds within worlds come to an end.

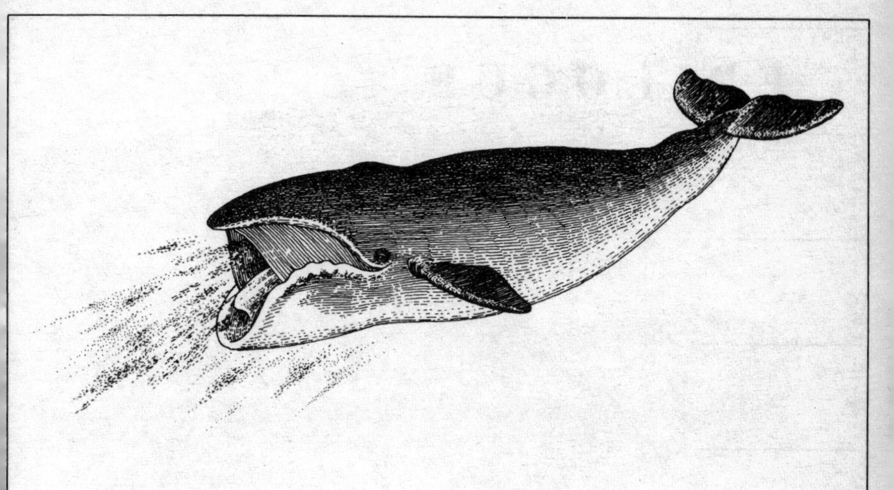

Right whales cruise along with their mouths open, filtering plankton as they go.

Rorquals feed by swimming up to a suitable patch of food, opening their mouths, and scooping up the whole lot, water, plankton, and fishes too.

EPILOGUE

THE BRITISH ARE well known for their sentimentality toward animals. The other day I heard a news item from England where an American wanted to swim the English Channel with his dog. The nation of animal lovers was outraged. An English veterinarian was interviewed, adding his professional weight to the popular outcry. The grueling fifteen-hour swim needed to make the twenty-two-mile crossing was hard enough for a human, but it would likely prove fatal for a dog. He was right too, but he did not go into the reasons. Here is another example of the effects of size differences on various species. The dog, being considerably smaller than the human, has a much larger area-to-volume ratio and would therefore suffer much greater heat losses. This story not only illustrates how scale differences affect all aspects of life, but also how these underlying principles are seldom acknowledged.

A wide range of size-related phenomena have been covered in this book, from the gliding mechanism of diatoms to the blood-pressure problems of dinosaurs. While diatoms and dinosaurs are pretty exotic creatures, most of the subjects encompassed by the individual chapters can be illustrated with examples drawn from everyday life. We began our journey by looking at the relationship between body size and metabolism. Small animals have higher metabolic rates than large ones and have to spend much of their time feeding to keep their metabolic flames burning brightly. As every parent knows, babies have to be fed every few hours, day and night, and it is only when they are bigger that they can manage on three meals a day. If a baby's regular feeding time is missed the parents will be reminded loudly and clearly. Adults, in contrast, often skip meals, sometimes without even noticing. We usually associate body mass with strength—big guys can lift heavier weights than small ones. But on the larger scale of things smaller animals are considerably stronger than large ones. You might find it hard work carrying a cooler full of goodies into the countryside, but consider the ease with which ants will carry away crumbs that are bigger than themselves.

The relation between body size and longevity is obvious to anyone who has ever had a pet. Many young hearts have been broken over the demise of a gerbil or a mouse, but cats and dogs often survive into their young owner's teens. Although a good correlation exists between body mass and longevity, there are many exceptions—like humans and horses, and parrots and ostriches—underscoring the fact that other factors are also involved. We saw that reproductive strategies are a key factor; animals such as mice, which breed young and produce many offspring, die young, whereas others, like humans, develop late, have fewer offspring, and live to a ripe old age.

I would not want to be an elephant—just watching them at the zoo makes me feel tired—but one of the advantages of being a giant is virtual immunity from predators. Elephants are the only giant land animals alive today, but the Mesozoic Era, the Age of Reptiles, abounded with dinosaurs that would have dwarfed an elephant. It is hard to imagine how an animal weighing as much as an airliner could possibly have walked on land, and most paleontologists used to believe that they spent their lives in lakes, buoyed up by the water.

However, the case that they did walk on land is a compelling one, and the most convincing evidence comes from trackways. The size and shape of the individual footprints show that they were made by sauropods and the clear impressions of fore and hind feet show that they were walking on all fours. Some particularly fine sauropod trackways from Texas even show where the soft mud they had been walking on had been squeezed into scrolls at the sides. If the beasts had been partly buoyed up their prints would have been shallow and the scrolls of mud would have been swirled away by the water. We saw how the sauropod's skeleton was engineered to withstand the forces of gravity, including the adoption of I-beam construction in their vertebrae. Although they walked on land, it is a matter of simple mechanics that they would have been unable to gambol and frolic like lambs the way they are often shown in some of the more imaginative depictions of Mesozoic life.

As a young lad I was very fond of my tortoise, but he was not a very demonstrative pet. My sister's pup, in contrast, was lots of fun and we could take her out for walks. There was little to choose between them in body mass, but the tortoise's brain was probably one-tenth that of the pup's, which helps explain the marked differences in their intellectual abilities. But the relation between relative brain size and intellect is not a simple one. Human brains, we saw, vary from about two to four pounds, but there does not appear to be any correlation with intellectual performance. As with many other parts of the body, including head size, brain size becomes progressively smaller as animals get bigger. Relative to its body mass, a lion's brain is about one tenth as big as a cat's, but a cat is no smarter than a lion. This helps explain why the sperm whale, which has the largest brain of any living animal (just over 20 pounds), is no smarter than the average mammal.

Although scaling problems of the brain can be frustratingly unpredictable, animals tend to be far more cooperative when it comes to flying and swimming. We know that fruit flies fly slowly enough to be caught by hand whereas house flies, which are about ten times bigger, fly much too fast. Some fliers are so small they make fruit flies look big. Not only do these minute fliers fly slowly, they are also at the mercy of the smallest air movements. A breeze can carry them miles away, while strong winds can whisk them across continents. At the other end of the spectrum is the albatross, whose eleven-foot wingspan can carry it across thousands of miles of ocean at speeds of fifty miles per hour. The squid, which compose the mainstay of the albatross's diet, are strong swimmers and can make their own way in the ocean. Most fish are the same, but the minutiae of marine life, the plankton, have only limited powers of swimming and must therefore drift with the flow. This strikes me as an attractive way of life, but most of the wanderers inevitably finish up inside the stomachs of larger creatures.

Diatoms and dinosaurs are poles apart in size, and worlds apart in life-styles, but are connected by common physical threads. When we unravel these threads we can begin to make sense of their diverse biological problems, and see the living world in a new and brighter light.

GLOSSARY

Allometric growth. Growth in which one part of the body changes size at a rate different from the rest of the body. The general form of the allometric equation is $y = ax^b$ where y = organ size, x = body size, b = the allometric exponent and a = the allometric coefficient. The logarithmic transformation of the equation is $\log y = \log a + b \log x$. When logarithmic values of y and x are plotted a straight-line graph is obtained whose slope is b and whose intercept with the y-axis is log a.

Altricial. Being hatched or born in a helpless state, where the youngster is often blind and unable to walk or feed itself, as in starlings and kittens.

Anterior. Toward the front, as in toward the head of an animal.

Arboreal. Living or spending much time in trees. The common squirrel is arboreal, as are many monkeys.

Arthropod. The largest phylum, or group of animals, which are characterized by their jointed legs. Included within the arthropods are spiders, crabs, and cockroaches.

Beam. A horizontal supporting structure, such as a steel girder in a building. A loaded beam experiences compressive stresses on its top surface and tensile stresses on its lower surface.

Benthic. Pertaining to the bottom of oceans and lakes. Corals, for example, are benthic organisms.

Bioluminescence. The phenomenon of living organisms emitting light. For example, fireflies (which are small beetles) are bioluminescent.

Biped. A two-legged animal, such as birds and humans.

Calorie. A calorie is the amount of heat required to raise the temperature of one gram of water by one Centigrade degree. The calories referred to in most books are kilocalories (1000 calories). For example, the calorific value of an ounce (28 grams) of butter is usually given as about 226 calories but these are actually kilocalories.

Carangiform. See *Thunniform.*

Carbohydrate. Compounds that contain carbon, hydrogen, and oxygen, where the ratio of hydrogen to oxygen is two to one as in water. Glucose and starch are examples.

Carnivora. An order of mammals that includes cats, bears, raccoons, otters, dogs, badgers, and ferrets. Most, but not all, of the order are meat-eaters.

Carnivores. Meat-eating animals.

Caudal. Pertaining to the tail, as in caudal vertebrae.

Cervical. Pertaining to the neck, as in cervical vertebrae.

Cetacea. The mammalian order that includes whales and dolphins.

Chloroplast. A small body within the cytoplasm of a plant cell that contains the pigment for carrying out photosynthesis.

Column. A vertical supporting structure, such as the vertical poles of a tent. A loaded column experiences mostly compressive stresses.

Compressive stress. The stress involved when things are pushed together. Standing on a brick would load it in compression.

Convergence. The phenomenon where animals or plants that are unrelated share similar features. For example, sharks and dolphins have similar torpedo-shaped bodies.

Cover slip. A thin plate or disk of glass that is placed over the top of a microscope slide to protect the specimen below. Cover slips have to be used when looking at plankton samples at high magnification to reduce the thickness of the film of water.

Diurnal. Pertaining to day. Diurnal animals are active during the daytime, in contrast to nocturnal ones.

Dorsal. At or near the back. The opposite of ventral.

Dorsal vertebrae. The vertebrae in the middle region of the vertebral column, between the pectoral and pelvic girdles.

Drag. The resistance that a body experiences when moving through a fluid. Drag has two major components: *pressure drag* (or form drag) attributable to the density of the fluid, and *friction drag* (or skin friction) attributable to its viscosity.

Echolocation. The act of locating objects by emitting high-pitched bursts of sound and detecting the direction of the reflected waves. Bats and odontocete whales use echolocation.

Encephalization quotient (EQ). A measure of the relative size of the brain that takes differential growth into account. It is numerically equal to the actual mass of the brain divided by the expected mass.

Energy. The product of force and distance. It has the same units as *work*.

Exoskeleton. The external skeleton of invertebrates like lobsters and ants.

Exponent. The power to which something is to be raised, that is, used as a factor. In the equation $y = x^2$, 2 is the exponent and indicates $x \times x$.

Fluid. A substance consisting of particles that move freely, in contrast to a solid where the component particles are fixed. Water and air are both fluids.

Flume. An apparatus in which water is circulated through a closed system of ducts so that models and aquatic animals can be tested in a continuous water flow—the aquatic equivalent of a wind-tunnel.

Frontal area. The area that a body presents to the fluid flow, measured perpendicular to the direction of flow. It can be visualized as the area of the shadow that the body would cast if a light source were substituted for the fluid flow.

Frugivorous. Fruit-eating.

Ganglion. An enlargement or body of nervous tissue containing a concentration of nerve cells.

Gliding. A style of flying on fixed wings in which no energy is extracted from the environment and height is lost.

Hemoglobin. The red pigment in blood that increases its capacity to carry oxygen.

Insectivorous. Insect-eating.

Joule. A unit of work or energy, abbreviated J. The work done by a force of one newton when it moves its point of application through a distance of one meter.

Kinetic energy. The energy of moving objects, which equals $\frac{1}{2}$ [mass $\times$ (velocity)^2].

Knot. One nautical mile per hour; 1.15 miles per hour (1.85 kilometers per hour).

Larva. A free-living immature stage of an animal's development.

Littoral. Pertaining to the seashore; the zone occurring between high and low tides.

Logarithm. The logarithm of a number is the exponent to which another number (the base, usually 10) must be raised to obtain the original number. For example, the logarithm of 1,000 is 3 because 10^3 is 1,000. In the days before pocket calculators tables of logarithms were used for doing multiplication and division.

Marsupial. Mammals like the kangaroo, in which the young are born at a very immature stage of development. Most marsupials have a pouch, the *marsupium*, in which the immature offspring complete their development.

Mass. A quantity of material, measured in grams, kilograms, pounds and the like. Mass is independent of gravity.

Mechanoreceptor. A biological device for converting mechanical impulses into nervous ones. There are, for example, stretch receptors in joints that convert changes in stress into nerve impulses.

Mesozoic Era. The time period extending from about 250 million years ago to 65 million years ago, when reptiles were the dominant vertebrates. Often referred to as the Age of Reptiles, or the Age of Dinosaurs, the era comprises three periods: the Triassic, Jurassic, and Cretaceous.

Metazoa. Organisms whose bodies are made up of many cells. We are metazoans, and so are all the familiar animals and plants. Compare *Protozoa*.

Micrometer. One thousandth of a millimeter, abbreviated μm (formerly called micron).

Monotremes. Mammals that lay eggs. There are only two living monotremes, the duck-billed platypus and the spiny anteater. Both live in Australia.

Myoglobin. A red oxygen-carrying pigment, similar to hemoglobin, found in muscles. Because it combines with oxygen at much lower concentrations than does hemoglobin, it functions as an important oxygen reservoir.

Nekton. A collective term for animals that have sufficiently strong swimming powers to be able to move at will, in contrast with plankton. Most fishes are nektonic.

Nerve ganglion. See *Ganglion*.

Newton. A unit of force, abbreviated N, which is the product of mass and acceleration. One newton is the amount of force needed to produce an acceleration of one meter per second per second in a mass of one kilogram.

Olfactory. Pertaining to the sense of smell.

Pascal. A unit of stress (or pressure), abbreviated Pa. One pascal is the amount of stress that occurs when one newton acts on one square meter.

Pectoral. Of the shoulder, as in the pectoral girdle.

Pelagic. Pertaining to the open sea.

Pelvic. Of the hips, as in the pelvic girdle.

Pennate (also pinnate). Having the appearance of a feather, as in pennate muscles and pennate diatoms.

Physiology. The science of the internal workings of living organisms.

Phytoplankton. A collective term for all the plants in the plankton; that is, those planktonic organisms that conduct photosynthesis.

Placental. Pertaining to the placenta.

Placental mammal. Mammals, such as our own species, where the fetus develops to full term within the uterus, supplied through the placenta with nutrients from the mother's blood. Most living mammals are placentals. Marsupial mammals, in contrast, are born in a very immature state and have to spend a long period of time developing inside the mother's pouch.

Plankter. An individual member of the plankton. The alternate term is *planktont*.

Plankton. A collective term for the plants and animals that drift with the currents in the upper layers of the ocean or lakes because of their limited or non-existent powers of locomotion. Most, but by no means all, members of the plankton are microscopic.

Planktont. See *Plankter.*

Power. The rate of doing work. When one joule of work is done in one second the power is one watt.

Precocial. Being hatched or born in a relatively advanced state where the individual is able to walk and see and is often able to feed itself, as in chickens and horses.

Primates. The mammalian order that contains lemurs, lorises, monkeys, apes, humans, and our fossil relatives.

Protozoa. Single-celled organisms, such as amoebas. Also called *acellular* because their bodies are not made of many cells. See *Metazoa.*

Quadruped. A four-legged animal, like a horse. Also called a *tetrapod*.

Reynolds number. An expression for the relative magnitudes of inertial and viscosity forces acting on a body moving through a fluid. The Reynolds number of such a body is obtained by multiplying the length and speed of the body by the density of the fluid and dividing by the viscosity of the fluid. Reynolds numbers are expressed without units.

Rodentia. An order of mammals, the rodents, containing mice, squirrels, beavers, porcupines, and rabbits.

Soaring. A style of flying on fixed wings in which lift is obtained by extracting energy from the environment rather than from muscles.

Stress. Force per unit area. Expressed in newtons per square meter, or pascals.

Taxonomic. Pertaining to the naming of living things.

Teleosts. A group of advanced bony fishes to which most modern species belong.

Tensile stress. The stress that is involved when things are pulled apart. Pulling on a rope would load it in tension.

Tetrapod. A four-legged animal, like a horse. Also called *quadruped*.

Thermal. A volume of moving air whose movements are caused by heating. In its simplest form a thermal is a column of warm air rising, say, from a rock heated by the sun.

Thunniform. A swimming mechanism in which the body is stiff and thrust is generated solely by the rapid oscillations of a crescentic tail. Named for the tunas, which exemplify this method of swimming. An alternate term is *carangiform.*

Ton. An English ton, called a "long ton" in the United States, is 2,240 pounds while a metric ton is 1,000 kilograms, or 2,200 pounds. As the two are so similar, I will not distinguish between them in this book.

Uropatagium. The membrane that stretches between the hind legs of a bat.

Ventral. Of or on the abdomen. The opposite of *dorsal.*

Watt. A unit of power; the rate of doing work. One watt is the amount of power used when one joule of work is performed in one second.

Weight. A quantity of material measured in the same units as mass but which varies with gravity. An object weighing one pound on earth would weigh considerably less on the moon.

Work. The product of force and distance. When one newton of force acts through a distance of one meter one joule of work has been performed.

Zooplankton. A collective term for all the animals in the plankton.

SOURCES OF FIGURES

All art is original unless noted below. Full citations of sources are given in the Further Reading list for each chapter.

CHAPTER 2

Page 22. Redrawn from Schmidt-Nielsen, 1979, *Animal Physiology.*

Page 23. Redrawn from Peters, 1983, *The Ecological Implications of Body Size.*

Page 24 (top). Redrawn from Heusner, 1982, "Energy metabolism and body size."

Page 24 (bottom). Redrawn from Heusner, 1991, "Size and power in mammals."

CHAPTER 3

Page 45. Redrawn from Heglund and Taylor, 1988, "Speed, stride frequency and energy cost per stride."

Page 46. Partly redrawn from Gray, 1953, *How Animals Move.*

CHAPTER 4

Page 56. Redrawn from Olshansky, Carnes, and Cassel, 1990, "In search of Methuselah."

Page 57. Redrawn from Lindstedt and Calder, 1976, "Body size and longevity in birds."

Page 63. Bill shapes redrawn from Grant and Grant, 1989, "Natural selection in a population of Darwin's finches."

CHAPTER 5

Page 79. Redrawn from R. Owen, 1866, *On the Anatomy of Vertebrates*, 3 vols. (London: Longmans, Green and Co.).

Pages 84, 85, 86. Partly based on Norman, 1985, *The Illustrated Encyclopedia of Dinosaurs*.

Page 87. Redrawn from a figure by O. C. Marsh, published in J. H. Ostrom and J. S. McIntosh, 1966, *Marsh's Dinosaurs* (New Haven: Yale University Press).

CHAPTER 6

Page 103. Redrawn from Jerison, 1973, *Evolution of the Brain and Intelligence*.

Page 104. Redrawn from Martin, 1981, "Relative brain size and basal metabolic rate in terrestrial vertebrates."

Page 114. Redrawn from Worthy and Hickie, 1986, "Relative brain size in marine animals."

Page 116. Redrawn from Roth et al., 1990, "Miniaturization in plethodontoid salamanders."

CHAPTER 8

Page 173. Redrawn from Hill and Smith, 1984, *Bats: A Natural History*.

Page 177. Redrawn from Padian and Rayner, 1993, "The wings of pterosaurs."

CHAPTER 9

Page 184. Based on the work of Ennos, 1988, "The importance of torsion in the design of insect wings."

Page 186. Adapted from Nachtigall, 1989, "Insect flight."

Page 190. *Left:* Modified from Weis-Fogh, 1973, "Quick estimates of flight fitness in hovering animals." *Right:* Redrawn from C. P. Clausen, 1940, *Entomophagous Insects* (New York: McGraw-Hill).

Page 194. Redrawn from T. Lewis, 1973, *Thrips—Their Biology, Ecology and Economic Importance* (London: Academic Press).

CHAPTER 10

Page 205. Redrawn from Edgar and Pickett-Heaps, 1984, "Diatom locomotion."

Pages 200, 203, 206, 208, 212, 215. Redrawn from various sources including G. E. Grove and R. C. Newell, 1963, *Marine Plankton* (London: Hutchinson Educational); and R. R. Kudo, 1960, *Protozoology* (Springfield: Charles C. Thomas).

Page 202. Modified from Lockhead, 1977, "Unsolved problems of interest in the locomotion of Crustacea."

Page 207. Redrawn from Vogel, 1981, *Life in Moving Fluids*.

Page 214. Modified from Berg, 1985, "Properties, problems and therapeutic progress" and 1988, "A physicist looks at bacterial chemotaxis."

CHAPTER II

Page 228. Redrawn from J. Gray, 1968, *Animal Locomotion* (London: Weidenfeld and Nicolson).

Page 240. Redrawn with modifications from Pivorunas, 1979, "The feeding mechanisms of baleen whales."

Page 242. Redrawn with modifications from Pivorunas, 1979, "Feeding mechanisms of baleen whales."

Page 243 (bottom). Redrawn from Sanderson and Wassersug, 1990, "Suspension-feeding vertebrates."

FURTHER READING

CHAPTER 1

Garland, T. 1983. "The relationship between maximal running speed and body mass in terrestrial mammals." *Journal of Zoology* 199:157–170.

Gibson, A. 1990. "Our chimp cousins get that much closer." *Science* 250:376.

Gould, S. J. 1974. "The origin and function of 'bizarre' structures: Antler size and skull size in the 'Irish Elk,' *Megaloceros giganteus*." *Evolution* 28: 191–220.

Harvey, P. H., and M. D. Pagel. 1991. *The Comparative Method in Evolutionary Biology*. Oxford: Oxford University Press.

Huxley, J. 1932. *Problems of Relative Growth*. London: Methuen.

Koop, B. F., M. Goodman, P. Xu, K. Chan, and J. L. Slightom. 1986. "Primate η-globin DNA sequences and man's place among the great apes." *Nature* 319:234–238.

Lasiewski, R. C. 1964. "Body temperatures, heart and breathing rate, and evaporative water loss in hummingbirds." *Physiological Zoology* 37: 212–223.

Promislow, D. E. L. 1993. On size and survival: Progress and pitfalls in the allometry of life span. *Journal of Gerontology* 48:B115–B123.

Schaller, G. B. 1972. *The Serengeti Lion*. Chicago: University of Chicago Press.

Schmidt-Nielsen, K. 1984. *Scaling: Why Is Animal Size So Important?* Cambridge: Cambridge University Press.

Spector, W. S., ed. 1956. *Handbook of Biological Data*. Philadelphia: W. B. Saunders.

CHAPTER 2

Bligh, J., and K. G. Johnson. 1973. "Glossary of terms for thermal physiology." *Journal of Applied Physiology* 35:941–961.

Block, B. A. 1991. "Evolutionary novelties: How fish have built a heater out of a muscle." *American Zoologist* 31:726–742.

Brattstrom, B.H. 1970. "Amphibia." In *Comparative Physiology of Thermoregulation*, ed. G.G. Whittow, vol. 1, pp. 135–166. New York: Academic Press.

Cabanac, M. 1987. "Glossary of terms for thermal physiology." *Pflügers Archiv* (*European Journal of Physiology*) 410:567–587.

Calder, W.A. 1984. *Size, Function, and Life History*. Cambridge, Mass.: Harvard University Press.

Carrier, D.R. 1987. "The evolution of locomotor stamina in tetrapods: Circumventing a mechanical constraint." *Paleobiology* 13:326–341.

Colbert, E.H. 1962. "The weights of dinosaurs." *American Museum Novitates* 2076:1–16.

Dawson, W.R., and J.W. Hudson. "Birds." In *Comparative Physiology of Thermoregulation*, ed. G.C. Whittow, vol. 1, pp. 223–310. New York: Academic Press.

Heusner, A.A. 1982. "Energy metabolism and body size. I. Is the 0.75 mass exponent of Kleiber's equation a statistical artifact?" *Respiratory Physiology* 48:1–12.

———. 1991. "Size and power in mammals." *Journal of Experimental Biology* 160:25–54.

Hill, J.R., and K.A. Rahimtulla. 1965. "Heat balance and the metabolic rate of new-born babies in relation to environmental temperatures and the effect of age and of weight on basal metabolic rate." *Journal of Physiology* 180:239–265.

Kleiber, M. 1932. "Body size and metabolism." *Hilgardia* 6:315–353.

MacArthur, R.A. 1979. "Seasonal patterns of body temperature and activity in free-ranging muskrats (*Ondatra zibethicus*)." *Canadian Journal of Zoology* 57:25–33.

McGowan, C. 1979. "Selection pressures for high body temperatures: Implications for dinosaurs." *Paleobiology* 5:285–295.

Mackay, R.S. 1964. "Galápagos tortoise and marine iguana deep body temperatures measured by radio telemetry." *Nature* 204:355–358.

Peters, R.H. 1983. *The Ecological Implications of Body Size*. Cambridge: Cambridge University Press.

Schmidt-Nielsen, K. 1979. *Animal Physiology: Adaptation and Environment*. Cambridge: Cambridge University Press.

———. 1984. *Scaling: Why Is Animal Size So Important?* Cambridge: Cambridge University Press.

Seymour, R. S., and D. F. Bradford. 1992. "Temperature regulation in the incubation mounds of the Australian Brush-turkey." *Condor* 94:134–150.

Slijper, E. J. 1962. *Whales*. London: Hutchinson.

Stone, G. N. 1993. "Endothermy in the solitary bee *Anthophora plumipes*: Independent measures of thermoregulatory ability, costs of warm-up and the role of body size." *Journal of Experimental Biology* 174:299–320.

Van Mierop, L. H. S., and S. M. Barnard. 1978. "Further observations on thermoregulation in the brooding female *Python molurus bivittatus* (Serpentes: Boidae)." *Copeia* 1978:615–621.

Whittow, G. C., and H. Tazawa. 1991. "The early development of thermoregulation in birds." *Physiological Zoology* 64:1371–1390.

Williams, T. M. 1990. "Heat transfer in elephants: thermal partitioning based on skin temperature profiles." *Journal of Experimental Biology* 222:235–245.

CHAPTER 3

Alexander, R. McN., V. A. Langman, and A. S. Jayes. 1977. "Fast locomotion of some African ungulates." *Journal of Zoology* 183:291–300.

Åstrand, P.-O., and A. Rodahl. 1986. *Textbook of Work Physiology*. New York: McGraw-Hill.

Biewener, A. A. 1989. "Scaling body support in mammals: Limb posture and muscle mechanics." *Science* 245:45–48.

Farley, C. T., and C. R. Taylor. 1991. "A mechanical trigger for the trot-gallop transition in horses." *Science* 253:306–308.

Garland, T. 1983. "The relationship between maximal running speed and body mass in terrestrial mammals." *Journal of Zoology* 199:157–170.

Gray, J. 1953. *How Animals Move*. Cambridge: Cambridge University Press.

Heglund, N. C., and C. R. Taylor. 1988. "Speed, stride frequency and energy cost per stride: How do they change with body size and gait?" *Journal of Experimental Biology* 138:301–318.

Heglund, N. C., C. R. Taylor, and T. A. McMahon. 1974. "Scaling stride frequency and gait to animal size: Mice to horses." *Science* 186:1112–1113.

Hill, A. V. 1950. "The dimensions of animals and their muscular dynamics." *Science Progress* 38:209–230.

McGowan, C. 1991. *Dinosaurs, Spitfires, and Sea Dragons*. Cambridge, Mass.: Harvard University Press.

Nadel, E. R. 1985. "Physiological adaptations to aerobic training." *American Scientist* 73:334–342.

Patak, A., and J. Baldwin. 1993. "Structural and metabolic characterization of the muscles used to power running in the Emu (*Dromaius novaehollandiae*), a giant flightless bird." *Journal of Experimental Biology* 175:233–249.

Pennycuick, C. J. 1992. *Newton Rules Biology*. Oxford: Oxford University Press.

Peters, S. E. 1989. "Structure and function in vertebrate skeletal muscle." *American Zoologist* 29:221–234.

Rome, L. C. 1992. "Scaling of muscle fibres and locomotion." *Journal of Experimental Biology* 168:243–252.

CHAPTER 4

Austad, S. N. 1992. "On the nature of aging." *Natural History* 2/92:25–56.

Beverton, R. J. H., and S. J. Holt. 1959. "A review of the lifespans and mortality rates of fish in nature, and their relation to growth and other physiological characteristics." In *The Lifespan of Animals*, ed. G. E. W. Wolstenholme and M. O'Connor, pp. 142–180. London: J. & A. Churchill.

Charnov, E. L. 1991. "Evolution of life history variation among female mammals." *Proceedings of the National Academy of Science* 88:1134–1137.

Comfort, A. 1956. *Ageing: The Biology of Senescence*. London: Routledge and Kegan Paul.

Finch, C. E. 1990. *Longevity, Senescence and the Genome*. Chicago: University of Chicago Press.

Flower, S. S. 1925. "Contributions to our knowledge of the duration of life in vertebrate animals. III. Reptiles." *Proceedings of the Zoological Society of London* 60:911–981.

———. 1937. "Further notes on the duration of life in animals. III. Reptiles." *Proceedings of the Zoological Society of London* 107:1–39.

———. "Further notes on the duration of life in animals. IV. Birds." *Proceedings of the Zoological Society of London* 108:195–235.

Germano, D. J. 1992. "Longevity and age-size relationships of populations of desert tortoises." *Copeia* 1992:367–374.

Gibbons, J. W. 1987. "Why do turtles live so long?" *BioScience* 37:262–269.

Grant, B. R., and P. R. Grant. 1989. "Natural selection in a population of Darwin's Finches." *American Naturalist* 133:377–393.

Harvey, P. H., M. D. Pagel, and J. A. Rees. 1991. "Mammalian metabolism and life histories." *American Naturalist* 137:556–566.

Harvey, P. H., and R. M. Zammuto. 1985. "Patterns of mortality and age at first reproduction in natural populations of mammals." *Nature* 315:319–320.

Hayflick, L. 1980. "The cell biology of human aging." *Scientific American* 242:58–65.

Lindstedt, S. L., and W. A. Calder. 1976. "Body size and longevity in birds." *Condor* 78:91–94.

———. 1981. "Body size, physiological time, and longevity of homeothermic animals." *Quarterly Review of Biology* 56:1–16.

Marsh, H., and T. Kasuya. 1986. "Evidence for reproductive senescence in female

cetaceans." In *Behaviour of Whales in Relation to Management*, ed. G. P. Donavan, pp. 57–75. Cambridge, Eng.: International Whaling Commission.
Olshansky, S. J., B. C. Carnes, and C. Cassel. 1990. "In search of Methuselah: Estimating the upper limits to human longevity." *Science* 250:634–640.
Pomeroy, D. 1990. "Why fly? The possible benefits for lower mortality." *Biological Journal of the Linnean Society* 40:53–65.
Promislow, D. E. L., and P. H. Harvey. 1990. "Living fast and dying young: A comparative analysis of life-history variation among mammals." *Journal of Zoology* 220:417–437.
———. 1993. "On size and survival: Progress and pitfalls in the allometry of life span." *Journal of Gerontology: Biological Sciences* 48:B115–B123.
Read, A. F., and P. H. Harvey. 1989. "Life history differences among the eutherian radiations." *Journal of Zoology* 219:329–353.
Rose, M. R. 1984. "The evolution of animal senescence." *Canadian Journal of Zoology* 62:1661–1667.
———. 1984. "Laboratory evolution of postponed senescence in *Drosophila melanogaster*." *Evolution* 38:1004–1010.
———. 1991. *Evolutionary Biology of Aging*. New York: Oxford University Press.
Sacher, G. A. 1959. "Relation of lifespan to brain weight and body weight in mammals." In *The Lifespan of Animals*, ed. G. E. W. Wolstenholme and M. O'Connor, pp. 115–141. London: J. & A. Churchill.
Sterrer, W. 1986. *Marine Fauna and Flora of Bermuda*. New York: John Wiley.

CHAPTER 5

Alexander, R. McN. 1985. "Mechanics of posture and gait of some large dinosaurs." *Zoological Journal of the Linnean Society* 83:1–25.
Bakker, R. T. 1978. "Dinosaur feeding behavior and the origin of flowering plants." *Nature* 274:661–663.
———. 1986. *The Dinosaur Heresies*. New York: William Morrow.
Bird, R. T. 1944. "Did *Brontosaurus* ever walk on land?" *Natural History* 53:60–67.
Brown, J. H., and B. A. Maurer. 1986. "Body size, ecological dominance and Cope's rule." *Nature* 324:248–250.
Brown, J. H., P. A. Marquet, and M. L. Taper. 1993. "Evolution of body size: Consequences of an energetic definition of fitness." *American Naturalist* 142:573–584.
Colbert, E. H. 1962. "The weights of dinosaurs." *American Museum Novitates* 2076:1–16.
Coombs, W. P. 1975. "Sauropod habits and habitats." *Palaeogeography, Palaeoclimatology, Palaeoecology* 17:1–33.
Demment, M. W., and P. J. Van Soest. 1985. "A nutritional explanation for body-size

patterns of ruminant and nonruminant herbivores." *American Naturalist* 125:641–672.

Dimery, N. J., R. McN. Alexander, and K. A. Deyst. 1985. "Mechanics of the ligamentum nuchae of some artiodactyls." *Journal of Zoology* 206:341–351.

Eltringham, S. K. 1982. *Elephants*. Poole, Eng.: Blandford Press.

Fortelius, M., and J. Kappelman. 1992. "New body mass estimates for the largest indricotheres." *Journal of Vertebrate Paleontology* 12:28A.

Hargens, A. R., R. W. Millard, K. Pettersson, and K. Johansen. 1987. "Gravitational haemodynamics and oedema prevention in the giraffe." *Nature* 329: 59–60.

Hicks, J. W., and H. S. Badeer. 1992. "Gravity and the circulation: 'open' vs. 'closed' systems." *American Journal of Physiology* 262:R725–R732.

Hohnke, L. A. 1973. "Haemodynamics in the Sauropoda." *Nature* 244:309–310.

Jensen, J. A. 1985. "Three new sauropod dinosaurs from the Upper Jurassic of Colorado." *Great Basin Naturalist* 45:697–709.

Martin, J. 1987. "Mobility and feeding of *Cetiosaurus* (Saurischia, Sauropoda)—Why the long neck?" In *Fourth Symposium on Mesozoic Ecosystems*, eds. P. J. Currie and E. H. Koster, pp. 150–155. Drumheller, Can.: Tyrell Museum of Palaeontology.

Norman, D. 1985. *The Illustrated Encyclopedia of Dinosaurs*. London: Salamander.

Rothschild, B. M. 1987. "Diffuse idiopathic skeletal hyperostosis as reflected in the paleontological record: Dinosaurs and early mammals." *Seminars in Arthritis and Rheumatism* 17:119–125.

Seymour, R. S. 1976. "Dinosaurs, endothermy and blood pressure." *Nature* 262:207–208.

Seymour, R. S., A. R. Hargens, and T. J. Pedley, 1993. "The heart works against gravity." *American Journal of Physiology* 265:R715–R720.

Seymour, R. S., and H. B. Lillywhite. 1976. "Blood pressure in snakes from different habitats." *Nature* 264:664–666.

Stanley, S. M. 1973. "An explanation for Cope's Rule." *Evolution* 27:1–26.

Toit, J. T. du, and N. Owen-Smith, 1989. "Body size, population metabolism, and habitat specialization among large African herbivores." *American Naturalist* 133:736–740.

Warren, J. V. 1974. "The physiology of the giraffe." *Scientific American* 231:96–105.

CHAPTER 6

Bennett, P. M., and P. H. Harvey. 1985. "Relative brain size and ecology in birds." *Journal of Zoology* 207:151–169.

———. 1985. "Brain size, development and metabolism in birds and mammals." *Journal of Zoology* 207:491–509.

Clayton, N. S., and J. R. Krebs. In preparation. "Hippocampal growth and attrition in birds affected by experience."

Clutton-Brock, T. H., and P. H. Harvey. 1979. "Comparison and adaptation." *Proceedings of the Royal Society of London* 205:547–565.

Cole, B. J. 1985. "Size and behavior in ants: Constraints on complexity." *Proceedings of the National Academy of Science* 82:8548–8551.

Eisenberg, J. F., and D. E. Wilson. 1978. "Relative brain size and feeding strategies in the Chiroptera." *Evolution* 32:740–751.

Gittleman, J. L. 1986. "Carnivore brain size, behavioral ecology, and phylogeny." *Journal of Mammalogy* 67:23–36.

———. 1991. "Carnivore olfactory bulb size: allometry, phylogeny and ecology." *Journal of Zoology* 225:253–272.

Gould, S. J. 1979. "Wide hats and narrow minds." *Natural History* 88:34–40.

Graham, A. J. W., and J. P. Hickie. 1986. "Relative brain size in marine mammals." *American Naturalist* 128:445–459.

Harvey, P. H., and J. R. Krebs. 1990. "Comparing brains." *Science* 249:140–146.

Healy, S., and T. Guilford. 1990. "Olfactory-bulb size and nocturnality in birds." *Evolution* 44:339–346.

Jerison, H. J. 1969. "Brain evolution and dinosaur brains." *American Naturalist* 934:575–588.

———. 1973. *Evolution of the Brain and Intelligence*. New York: Academic Press.

Krebs, J. R., D. F. Sherry, S. D. Healy, V. H. Perry, and A. L. Vaccarino. 1989. "Hippocampal specialization of food-storing birds." *Proceedings of the National Academy of Science* 86:1388–1392.

Martin, R. D. 1981. "Relative brain size and basal metabolic rate in terrestrial vertebrates." *Nature* 293:57–60.

Meier, P. T. 1983. "Relative brain size within the North American Sciuridae." *Journal of Mammalogy* 64:642–647.

Mobbs, P. G. 1985. "Brain structure." In *Comprehensive Insect Physiology, Biochemistry, and Pharmacology*, ed. G. A. Kerkut and L. I. Gilbert, pp. 299–370. Volume 5. Oxford: Pergamon Press.

Pagel, M. D., and P. H. Harvey. 1990. "Diversity in the brain sizes of newborn mammals." *BioScience* 40:116–122.

Roth, G., B. Rottluff, W. Grunwald, J. Hanken, and R. Linke. 1990. "Miniaturization in plethodontid salamanders (Caudata: Plethodontidae) and its consequences for the brain and visual system." *Biological Journal of the Linnean Society* 40:165–190.

Russell, A., ed. 1988. *Guinness Book of World Records*. New York: Sterling.

Schmidt, J. M., and J. J. B. Smith. 1987. "The measurement of exposed host volume by the parasitoid wasp *Trichogramma minutum* and the effects of wasp size." *Canadian Journal of Zoology* 65:2837–2845.

———. 1989. "Host examination walk and oviposition site selection of *Trichogramma minutum*: Studies on spherical hosts." *Journal of Insect Behavior* 2:143–171.

Sherry, D. F., L. F. Jacobs, and S. J. C. Gaulin. 1992. "Spatial memory and adaptive specialization of the hippocampus." *Trends in Neuroscience* 15:298–303.

Shettleworth, S. J. 1990. "Spatial memory in food-storing birds." *Philosophical Transactions of the Royal Society* 329:143–151.

Willerman, L., R. S. Schultz, J. N. Rutledge, and E. D. Bigler. 1991. "*In vivo* brain size and intelligence." *Intelligence* 15:223–228.

Worthy, G. A. J., and J. P. Hickie. 1986. "Relative brain size in marine animals." *American Naturalist* 128:445–459.

CHAPTER 7

Blake, R. W. 1983. *Fish Locomotion*. Cambridge: Cambridge University Press.

Green, W. 1975. *Famous Fighters of the Second World War*. London: MacDonald.

Harris, J. E. 1936. "The role of the fins in the equilibrium of the swimming fish. I. Wind-tunnel tests on a model of *Mustelus canis* (Mitchill)." *Journal of Experimental Biology* 23:476–493.

Kermode, A. C. 1972. *Mechanics of Flight*. London: Pitman Books.

McGowan, C. 1991. *Dinosaurs, Spitfires, and Sea Dragons*. Cambridge, Mass.: Harvard University Press.

CHAPTER 8

Baudinette, R. V., and K. Schmidt-Nielsen. 1974. "Energy cost of gliding flight in herring gulls." *Nature* 248:83–84.

Bramwell, C. D., and G. R. Whitfield. 1974. "Biomechanics of *Pteranodon*." *Philosophical Transactions of the Royal Society of London* 267:503–581.

Brower, J. C. 1983. "The aerodynamics of *Pteranodon* and *Nyctosaurus*, two large pterosaurs from the Upper Cretaceous of Kansas." *Journal of Vertebrate Paleontology* 3:84–124.

Brown, R. E., and M. R. Fedde. 1993. "Airflow sensors in the avian wing." *Journal of Experimental Biology* 179:13–30.

Butler, P. J. 1991. "Exercise in birds." *Journal of Experimental Biology* 160:233–262.

Costa, D. P., and P. A. Prince. 1987. "Foraging energetics of grey-headed albatrosses *Diomedea chrysostoma* at Bird Island, South Georgia." *Ibis* 129:149–158.

Darwin, C. 1896. *Journal of Researches into the Natural History and Geology of the Countries Visited during the Voyage of H.M.S. Beagle Round the World, under the Command of Capt. FitzRoy, R.N.* New York: D. Appleton.

Ellington, C. P. 1991. "Limitations on animal flight performance." *Journal of Experimental Biology* 160:71–91.

Fenton, M. B. 1983. *Just Bats*. Toronto: University of Toronto Press.

Greenewalt, C. H. 1961. *Hummingbirds*. New York: Doubleday.

———. 1962. "Dimensional relationships for flying animals." *Smithsonian Miscellaneous Collections* 144:1–46.

———. 1975. "Could pterosaurs fly?" *Science* 188:676–677.

Guston, B. 1977. *Modern Military Aircraft*. London: Salamander.

Hankin, E. H. 1910. *Animal Flight*. London: Iliffe and Sons, Ltd.

Hanson, J. R. 1991. "Bigger: The quest for size." In *Milestones of Aviation*, ed. J. T. Greenwood, pp. 150–221. New York: Crescent Books.

Hill, J. E., and J. D. Smith. 1984. *Bats: A Natural History*. Austin: University of Texas Press.

Jouventin, P., and H. Weimerskirch. 1990. "Satellite tracking of Wandering albatrosses." *Nature* 343:746–748.

Ker, R. F., M. B. Bennett, S. R. Bibby, R. C. Kester, and R. McN. Alexander. 1987. "The spring in the arch of the human foot." *Nature* 325:147–149.

Lachmann, G. V. 1964. "Sir Frederick Handley Page, the man and his work." *Journal of the Royal Aeronautical Society* 68:433–452.

Langston, W., Jr. 1981. "Pterosaurs." *Scientific American* 244:122–136.

Lawson, D. A. 1975. "Pterosaur from the latest Cretaceous of West Texas: Discovery of the largest flying creature." *Science* 187:947–948.

Lighthill, J. 1975. "Aerodynamic aspects of animal flight." In *Swimming and Flying in Nature*, eds. T. Y.-T. Wu, C. J. Brokaw, and C. Brennen, vol. 2, pp. 423–491. New York: Plenum Press.

McGowan, C. 1991. *Dinosaurs, Spitfires, and Sea Dragons*. Cambridge, Mass.: Harvard University Press.

Marden, J. H. 1987. "Maximum lift production during takeoff in flying animals." *Journal of Experimental Biology* 130:235–258.

———. 1990. "Maximum load-lifting and induced power output of Harris' hawks are general functions of flight muscle mass." *Journal of Experimental Biology* 149:511–514.

———. In press. "From damselflies to dinosaurs: How burst and sustained flight performance scale with size." *American Journal of Physiology*.

Oehme, H. 1977. "On the aerodynamics of separated primaries in the avian wing." In *Scale Effects in Animal Locomotion*, ed. T. J. Pedley, pp. 479–495. London: Academic Press.

Padian, K. 1983. "A functional analysis of flying and walking in pterosaurs." *Paleobiology* 9:218–239.

Padian, K., and J. M. V. Rayner. 1993. "The wings of pterosaurs." *American Journal of Science* 293-A:91–166.

Pennycuick, C. J. 1972a. "Soaring behavior and performance of some East African birds, observed from a motor-glider." *Ibis* 114:178–218.

———. 1972b. *Animal Flight*. London: Edward Arnold.

———. 1982. "The flight of petrels and albatrosses (Procellariiformes), observed in South Georgia and its vicinity." *Philosophical Transactions of the Royal Society of London* 300:75–106.

———. 1983. "Thermal soaring compared in three dissimilar tropical bird species, *Fregata magnificens, Pelecanus occidentalis* and *Coragyps atratus*." *Journal of Experimental Biology* 102:307–325.

Swartz, S. M., M. B. Bennett, and D. R. Carrier. 1992. "Wing bone stresses in free flying bats and the evolution of skeletal design for flight." *Nature* 359:726–729.

Tucker, V. A. 1977. "Scaling in avian flight." In *Scale Effects in Animal Locomotion*, ed. T. J. Pedley, pp. 497–509. London: Academic Press.

———. 1993. "Gliding birds: Reduction of induced drag by wing-tip slots between the primary feathers." *Journal of Experimental Biology* 180:285–310.

Unwin, D. M. 1987. "Pterosaur locomotion: joggers or waddlers?" *Nature* 327:13–14.

Wellnhofer, P. 1991. *The Illustrated Encyclopedia of Pterosaurs*. London: Salamander.

Wells, D. J. 1993. "Muscle performance in hovering hummingbirds." *Journal of Experimental Biology* 178:39–57.

White, G. 1880. *The Natural History and Antiquities of Selborne*. New edition. London: George Routledge and Sons.

CHAPTER 9

Ellington, C. P. 1975. "Non-steady aerodynamics of the flight of *Encarsia formosa*." In *Swimming and Flying in Nature*, eds. T. Y.-T. Wu, C. J. Brokaw, and C. Brennen, vol. 2, pp. 423–491. New York: Plenum Press.

———. 1984. "The aerodynamics of hovering insect flight. I. The quasi-steady analysis. II. Morphological parameters. III. Kinematics. IV. Aerodynamic mechanisms." *Philosophical Transactions of the Royal Society of London* B 305:1–15, 17–40, 41–78, 79–113.

Ennos, A. R. 1988. "The importance of torsion in the design of insect wings." *Journal of Experimental Biology* 140:137–160.

Gibo, D. L., and M. J. Pallett. 1979. "Soaring flight of monarch butterflies, *Danaus plexippus* (Lepidoptera: Danaidae), during the late summer migration in southern Ontario." *Canadian Journal of Zoology* 57:1393–1401.

Glick, P. A. 1939. "The distribution of insects, spiders, and mites in the air." *Technical Bulletin United States Department of Agriculture* 673:1–150.

Hacklinger, M. 1964. "Theoretical and experimental investigation of indoor flying models." *Journal of the Royal Aeronautical Society* 68:728–734.

Johnson, C. G. 1969. *Migration and Dispersal of Insects by Flight*. London: Methuen.

Kuethe, A. M. 1975. "On the mechanics of flight of small insects." In *Swimming and*

Flying in Nature, eds. T. Y.-T. Wu, C. J. Brokaw, and C. Brennen, vol. 2, pp. 803–813. New York: Plenum Press.

Lighthill, J. 1975. "Aerodynamic aspects of animal flight." In *Swimming and Flying in Nature*, eds. T. Y.-T. Wu, C. J. Brokaw, and C. Brennen, vol. 2, pp. 423–491. New York: Plenum Press.

Marden, J. H., and P. Chai. 1991. "Aerial predation and butterfly design: How palatability, mimicry, and the need for evasive flight constrain mass allocation." *American Naturalist* 138:15–36.

Nachtigall, W. 1989. "Insect flight." In *Insect Flight*, eds. G. J. Goldsworth and C. H. Wheeler, pp. 1–29. Boca Raton, Fla.: CRC Press.

Schmidt, J. M., and J. J. B. Smith. 1985. "The ultrastructure of the wings and the external sensory morphology of the thorax in female *Trichogramma minutum* Riley (Hymenoptera: Chalcidoidea: Trichogrammatidae)." *Proceedings of the Royal Society of London* 224:287–313.

Vogel, S. 1981. *Life in Moving Fluids*. Princeton: Princeton University Press.

Weis-Fogh, T. 1973. "Quick estimates of flight fitness in hovering animals, including novel mechanisms for lift production." *Journal of Experimental Biology* 59:169–230.

———. 1975. "Flapping flight and power in birds and insects, conventional and novel mechanisms." In *Swimming and Flying in Nature*, eds. T. Y.-T. Wu, C. J. Brokaw, and C. Brennen, vol. 2, pp. 729–762. New York: Plenum Press.

Yoshimoto, C. M., and J. L. Gressitt. 1960. "Trapping of air-borne insects on ships on the Pacific (part 3)." *Pacific Insects* 2:239–243.

———. 1961. "Trapping of air-borne insects on ships on the Pacific (part 4)." *Pacific Insects* 3:556–558.

CHAPTER 10

Berg, H. C. 1985. "Properties, problems and therapeutic progress: Flagellar motility." *Roche Seminars on Bacteria* no. 1 (Reprinted 1990): 1–23.

———. 1988. "A physicist looks at bacterial chemotaxis." *Cold Spring Harbor Symposia on Quantitative Biology* 53:1–10.

———. 1990. "Bacterial microprocessing." *Cold Spring Harbor Symposia on Quantitative Biology* 55:539–545.

Berg, H. C., and R. A. Anderson. 1973. "Bacteria swim by rotating their flagellar filaments." *Nature* 245:380–382.

Brokaw, C. J. 1975. "Mechanisms of movement in flagella and cilia." In *Swimming and Flying in Nature*, eds. T. Y.-T. Wu, C. J. Brokaw, and C. Brennen, vol. 1, pp. 89–126. New York: Plenum Press.

De Mott, W. R., and M. D. Watson. 1991. "Remote detection of algae by copepods: response to algal size, odors and motility." *Journal of Plankton Research* 13:1203–1222.

Dial, K. P., and A. A. Biewener. 1993. "Pectoralis muscle force and power output during different modes of flight in pigeons (*Columba livia*)." *Journal of Experimental Biology* 176:31–54.

Edgar, L. A., and J. D. Pickett-Heaps. 1984. "Diatom locomotion." In *Progress in Phycological Research*, eds. Round and Chapman, vol. 3, pp. 48–88. Bristol: Bio Press Ltd.

Lockhead, J. H. 1977. "Unsolved problems of interest in the locomotion of Crustacea." In *Scale Effects in Animal Locomotion*, ed. T. J. Pedley, pp. 257–268. London: Academic Press.

Lowndes, A. G. 1935. "The swimming and feeding of certain calanoid copepods." *Proceedings of the Zoological Society of London* 1935:687–715.

Purcell, E. M. 1977. "Life at low Reynolds number." *American Journal of Physics* 45:3–11.

Sleigh, M. A., and J. R. Blake. 1977. "Methods of ciliary propulsion and their size limitations." In *Scale Effects in Animal Locomotion*, ed. T. J. Pedley, pp. 243–256. London: Academic Press.

Strickler, J. R. 1975. "Swimming of planktonic *Cyclops* species (Copepoda, Crustacea): Pattern, movements and their control." In *Swimming and Flying in Nature*, eds. T. Y.-T. Wu, C. J. Brokaw, and C. Brennen, vol. 2, pp. 599–613. New York: Plenum Press.

Vogel, S. 1981. *Life in Moving Fluids*. Princeton: Princeton University Press.

CHAPTER 11

Blake, R. W. 1983. *Fish Locomotion*. Cambridge: Cambridge University Press.

Block, B. A., and D. Booth. 1992. "Direct measurement of swimming speeds and depth of blue marlin." *Journal of Experimental Biology* 166:267–284.

Bone, Q. 1975. "Muscular and energetic aspects of fish swimming." In *Swimming and Flying in Nature*, eds. T. Y.-T. Wu, C. J. Brokaw, and C. Brennen, vol. 2, pp. 493–528. New York: Plenum Press.

Carey, F. G., and B. H. Robinson. 1981. "Daily patterns in the activities of swordfish, *Xiphias gladius*, observed by acoustic telemetry." *Fisheries Bulletin* 79:277–292.

Carey, F. G., and J. V. Scharold. 1990. "Movements of blue sharks (*Prionace glauca*) in depth and course." *Marine Biology* 106:329–342.

Dewar, H., and J. B. Graham. 1968. *Animal Locomotion*. London: Weidenfeld and Nicolson.

———. In press. "Studies of tropical tuna swimming performance in a large water tunnel. I: Energetics." *Journal of Experimental Biology*.

Goode, G. B. 1883. "Materials for a history of the sword-fish." *United States Commercial Fish Fisheries, Part 8, Report of the Fisheries Commission* 1883:287–394.

Graham, J. B., H. Dewar, N. C. Lai, W. R. Lowell, and S. M. Arce. 1990. "Aspects of shark swimming performance determined using a large water tunnel." *Journal of Experimental Biology* 151:175–192.

Gray, J. 1936. "Studies in animal locomotion." *Journal of Experimental Biology* 13:192–199.

Hoyt, J. W. 1975. "Hydrodynamic drag reduction due to fish slimes." In *Swimming and Flying in Nature*, eds. T. Y.-T. Wu, C. J. Brokaw, and C. Brennen, vol. 2, pp. 653–672. New York: Plenum Press.

Lang, T. G. 1975. "Speed, power, and drag measurements of dolphins." In *Swimming and Flying in Nature*, eds. T. Y.-T. Wu, C. J. Brokaw, and C. Brennen, vol. 2, pp. 553–572. New York: Plenum Press.

McGowan, C. 1992. "The ichthyosaurian tail: Sharks do not provide an appropriate analogue." *Palaeontology* 35:555–570.

Pivorunas, A. 1979. "The feeding mechanisms of baleen whales." *American Scientist* 67:432–440.

Sanderson, S. L., and R. Wassersug. 1990. "Suspension-feeding vertebrates." *Scientific American* 262:96–101.

Thomson, K. S., and D. E. Simanek. 1977. "Body form and locomotion in sharks." *American Zoologist* 17:343–354.

Van Dam, C. P. 1987. "Efficiency characteristics of crescent-shaped wings and caudal fins." *Nature* 325:435–437.

Wardle, C. S. 1977. "Effects of size on the swimming speeds of fish." In *Scale Effects in Animal Locomotion*, ed. T. J. Pedley, pp. 299–313. London: Academic Press.

Webb, P. W. 1982. "Locomotor patterns in the evolution of actinopterygian fishes." *American Zoologist* 22:329–342.

———. 1984. "Form and function in fish swimming." *Scientific American* 251:72–82.

———. 1988. "Simple physical principles and vertebrate aquatic locomotion." *American Zoologist* 28:709–725.

Webb, P. W., and R. W. Blake. 1985. "Swimming." In *Functional Vertebrate Morphology*, ed. M. Hildebrand, D. M. Bramble, K. F. Liem, and D. W. Wake, pp. 110–128. Cambridge, Mass.: Harvard University Press.

Webb, P. W., and V. de Buffrénil. 1991. "Locomotion in the Biology of Large Aquatic Vertebrates." *Transactions of the American Fisheries Society* 119:629–641.

INDEX

READ MORE IN PENGUIN

In every corner of the world, on every subject under the sun, Penguin represents quality and variety – the very best in publishing today.

For complete information about books available from Penguin – including Puffins, Penguin Classics and Arkana – and how to order them, write to us at the appropriate address below. Please note that for copyright reasons the selection of books varies from country to country.

In the United Kingdom: Please write to *Dept. EP, Penguin Books Ltd, Bath Road, Harmondsworth, West Drayton, Middlesex UB7 ODA*

In the United States: Please write to *Consumer Sales, Penguin Putnam Inc., P.O. Box 12289 Dept. B, Newark, New Jersey 07101-5289.* VISA and MasterCard holders call 1-800-788-6262 to order Penguin titles

In Canada: Please write to *Penguin Books Canada Ltd, 10 Alcorn Avenue, Suite 300, Toronto, Ontario M4V 3B2*

In Australia: Please write to *Penguin Books Australia Ltd, P.O. Box 257, Ringwood, Victoria 3134*

In New Zealand: Please write to *Penguin Books (NZ) Ltd, Private Bag 102902, North Shore Mail Centre, Auckland 10*

In India: Please write to *Penguin Books India Pvt Ltd, 11 Community Centre, Panchsheel Park, New Delhi 110017*

In the Netherlands: Please write to *Penguin Books Netherlands bv, Postbus 3507, NL-1001 AH Amsterdam*

In Germany: Please write to *Penguin Books Deutschland GmbH, Metzlerstrasse 26, 60594 Frankfurt am Main*

In Spain: Please write to *Penguin Books S. A., Bravo Murillo 19, 1° B, 28015 Madrid*

In Italy: Please write to *Penguin Italia s.r.l., Via Benedetto Croce 2, 20094 Corsico, Milano*

In France: Please write to *Penguin France, Le Carré Wilson, 62 rue Benjamin Baillaud, 31500 Toulouse*

In Japan: Please write to *Penguin Books Japan Ltd, Kaneko Building, 2-3-25 Koraku, Bunkyo-Ku, Tokyo 112*

In South Africa: Please write to *Penguin Books South Africa (Pty) Ltd, Private Bag X14, Parkview, 2122 Johannesburg*

READ MORE IN PENGUIN

SCIENCE AND MATHEMATICS

Six Easy Pieces Richard P. Feynman

Drawn from his celebrated and landmark text *Lectures on Physics*, this collection of essays introduces the essentials of physics to the general reader. 'If one book was all that could be passed on to the next generation of scientists it would undoubtedly have to be *Six Easy Pieces*' John Gribbin, *New Scientist*

A Mathematician Reads the Newspapers John Allen Paulos

In this book, John Allen Paulos continues his liberating campaign against mathematical illiteracy. 'Mathematics is all around you. And it's a great defence against the sharks, cowboys and liars who want your vote, your money or your life' Ian Stewart

Dinosaur in a Haystack Stephen Jay Gould

'Today we have many outstanding science writers . . . but, whether he is writing about pandas or Jurassic Park, none grabs you so powerfully and personally as Stephen Jay Gould . . . he is not merely a pleasure but an education and a chronicler of the times' *Observer*

Does God Play Dice? Ian Stewart

As Ian Stewart shows in this stimulating and accessible account, the key to this unpredictable world can be found in the concept of chaos, one of the most exciting breakthroughs in recent decades. 'A fine introduction to a complex subject' *Daily Telegraph*

About Time Paul Davies

'With his usual clarity and flair, Davies argues that time in the twentieth century is Einstein's time and sets out on a fascinating discussion of why Einstein's can't be the last word on the subject' *Independent on Sunday*